Advanced Studies
Mobile Research Center Bremen

Edited by
A. Förster,
C. Görg,
O. Herzog,
M. Lawo,
H. Witt,
Bremen, Germany

Das Mobile Research Center Bremen (MRC) im Technologie-Zentrum Informatik und Informationstechnik (TZI) der Universität Bremen erforscht, entwickelt und erprobt in enger Zusammenarbeit mit der Wirtschaft mobile Informatik-, Informations- und Kommunikationstechnologien. Als Forschungs- und Transferinstitut des Landes Bremen vernetzt und koordiniert das MRC hochschulübergreifend eine Vielzahl von interdisziplinären Arbeitsgruppen, die sich mit der Entwicklung und Anwendung mobiler Lösungen beschäftigen. Die Reihe „Advanced Studies" präsentiert ausgewählte hervorragende Arbeitsergebnisse aus der Forschungstätigkeit der Mitglieder des MRC.

In close collaboration with the industry, the Mobile Research Center Bremen (MRC), a division of the Center for Computing and Communication Technologies (TZI) of the University of Bremen, investigates, develops and tests mobile computing, information and communication technologies. This research cluster of the state of Bremen links and coordinates interdisciplinary research teams from different universities and institutions, which are concerned with the development and application of mobile solutions. The series "Advanced Studies" presents a selection of outstanding results of MRC's research projects.

Edited by
Prof. Dr. Anna Förster
Kommunikationsnetze - FB 1
Bremen, Germany

Prof. Dr. Carmelita Görg
Kommunikationsnetze - FB 1
Bremen, Germany

Prof. Dr. Otthein Herzog
Centre for Computers and
Communication Technology
Bremen, Germany

Prof. Dr. Michael Lawo
Mobile Research Center (MRC)
Bremen, Germany

Prof. Dr. Hendrik Witt
Schuppen 2
Bremen, Germany

Max Gath

Optimizing Transport Logistics Processes with Multiagent Planning and Control

With a foreword by Prof. Dr. Otthein Herzog

Max Gath
Bremen, Germany

Dissertation Universität Bremen, Germany, 2015

Advanced Studies Mobile Research Center Bremen
ISBN 978-3-658-14002-1   ISBN 978-3-658-14003-8 (eBook)
DOI 10.1007/978-3-658-14003-8

Library of Congress Control Number: 2016939112

Springer Vieweg
© Springer Fachmedien Wiesbaden 2016

Printed on acid-free paper

This Springer Vieweg imprint is published by Springer Nature
The registered company is Springer Fachmedien Wiesbaden GmbH

Foreword

There is huge evidence that the logistics of transportation processes in cities needs much more attention than some years ago. There are not only the worldwide value creation networks anymore which require more efficient logistics based on concepts such as Industry 4.0, Made in China/China 2025, Industrial Internet (and more). In addition, we are already talking about the "Amazon" or "Jack Ma" societies where many goods are ordered through the Internet and must be delivered as quickly as possible, often on the same day or even within a few hours, in smaller delivery batches than ever before. The effects on our cities are clearly observable: delivery trucks all over the place, often blocking the traffic flow. It does not need to be mentioned that these new logistics services should be offered at very competitive cost and obeying all ecological requirements in spite of their obvious inefficiency. Moreover, a new "Logistics 4.0" could be an excellent application area for new services, e.g., for commercial logistics service offerings in the cloud as SaaS.

If we analyze the current state-of-the-art of logistics processes in optimization, planning and controlling the transportation within cities or regions, it becomes obvious that the current delivery processes offer many opportunities to improve their efficiency: delivery planning is still done so much ahead of time that it cannot be varied anymore if conditions change. And which plan could be valid throughout a full day given the plethora of disturbances in city traffic? Additionally, because of the urgency, the full capacity of delivery trucks is not utilized, and also often the delivery routes are already obsolete in the morning. This inefficiency is even exacerbated if delivery is combined with pickup given the additional (often unreliable) time windows to be scheduled.

These observations should suffice to understand that we have to solve a host of problems if logistics cost in Western countries should not increase beyond 10% of the GDP. Dr. Gath has accepted this challenge and defined the following goals in order to contribute to a solution:

- lowered cost through efficient, flexible, robust and adaptive planning,

- better protection of the environment through a better capacity utilization of truck fleets, through near-optimal delivery/pickup tours, and through improved route planning,

- better service and improved satisfaction of the delivery and pickup customers by satisfying all well-known general logistics requirements,

- development of a software SaaS architecture suited to future commercial purposes.

In addition, Dr. Gath has deepened this approach by aiming at most general problem solutions by concentrating first on basic concepts and only then attacking the implementation of optimization, planning, and control of the logistics processes. These concepts include auction protocols for Multiagent Systems as a technology supporting autonomous processes, and the most important basic functions such as finding the most efficient implementation of shortest paths for optimal routes and also for efficient algorithms for the solution of the Traveling Salesman Problem. Therefore, he could show that his solution is competitive or even outperforms other multiagent-based approaches compared to standard benchmarks, but even more important, using real data from the field he showed that his solution generates better and more flexible plans than two existing systems actually used in logistic enterprises. This must have been a convincing argument for one of the application partners as they incorporated his new system into their operational system to reap the benefits of the joint work – which usually does not happen very often in industry-university co-operations, but demonstrates nicely the merits of the solution.

It is very fortunate that this successful endeavor will be now available to the interested public in order to mitigate the described negative effects of current transportation logistics. My wish would be that these new results will be used much more often so that the needed economic and the ecological benefit of better transportation logistics can be significantly increased.

Otthein Herzog

Acknowledgements

First, I would like to thank my advisor Prof. Dr. Otthein Herzog for his notable support during my research. He enabled me to develop new approaches which can be transferred from research to real-world processes in logistics. I very much appreciate the time, valuable remarks, and comments he dedicated to my projects.

I also wish to express my sincere thanks to Prof. Dr. Winfried Lamersdorf, who kindly agreed to review this thesis. Furthermore, I would like to thank Prof. Dr. Stefan Edelkamp for his advice and for offering his broad knowledge of artificial intelligence and algorithm engineering. Also, Prof. Dr. Michael Lawo provided valuable help and remarks from the perspective of management.

My sincere thanks also go to Jens Engelmann, Reinhard Riefflin, Sven Bünger, and Tobias Kohsman from Hellmann Worldwide Logistics. They gave me the unique chance to investigate and analyze real-world processes in groupage traffic by sharing their long-time expert knowledge and by participating in their departments in the Bremen Hellmann office.

Likewise, I thank Mateusz Juraszek, Jacek Becela, Michael Löhr, Thomas Bluth, Dirk Reiche, and Philipp Walz from tiramizoo for a great cooperation during the last few years. They enabled me to extend approaches in order to apply them in the challenging domain of courier, express, and parcel services with same-day deliveries.

Especially, I also would like to thank the experts in logistics and artificial intelligence Dr. Arne Schuld and Dr. Jan D. Gehrke from Aimpulse Intelligent Systems for their collaboration. They provided valuable support during all my research.

Moreover, a big thank you to all my colleagues at the research group Artificial Intelligence at the Center of Computing and Communication

Technologies (TZI), Universität Bremen, as well as the LogDynamics team for stimulating discussions. I am especially grateful to Christoph Greulich, Tobias Warden, Dr. Jan Ole Bernd, Florian Pantke, Malte Humann, and David Zastrau for the teamwork on the PlaSMA project and for all the fun we have had in the last few years.

Finally, I wish to thank Dr. Thomas Wagner who was a great supervisor of my diploma thesis and who was the one who introduced me to multiagent systems in the first place.

Max Gath

Contents

List of Figures

List of Tables

1 Introduction

In the next few years, manufacturing and production processes will be changing significantly. Customers expect high-quality, custom-made products within the shortest time and at the lowest cost. New technologies, globalization, the dismantling of trade barriers, and additional free trade arrangements enable customers to buy their products at nearly every point on earth from the company which best satisfies their needs. As a result, client loyalty is decreasing and the change from buyer to customer markets continues.

Therefore, demanding customer requirements and growing cost pressure in the global market force enterprises to extend their product-ranges and to increase their service level, their efficiency, and their reliability at lower overall costs. In addition, the product life-cycle must be significantly shortened in order to reduce the time-to-market which is essential for the implementation of mass-customization in manufacturing processes. This is a challenge, especially in high-wage countries with highest social and environmental standards.

In order to achieve these goals, the fourth industrial revolution has been started with the development of Industry 4.0 (Kagemann, Wahlster, and Helbig, 2013), which is similar to the so-called Industrial Internet in the USA. In the first industrial revolution, the introduction of steam engines lead to mechanical production systems which for the first time enabled mass-production of uniformed goods at the end of the 18th century. These manufacturing processes were enhanced by electricity driven production plants and the division of labor in the second industrial revolution at the beginning of the 20th century. In the third industrial revolution, information and communication technologies have been applied for the automation of manufacturing processes since around 1970.

Today, the goal of the fourth industrial revolution is the mass-customization of processes to efficiently produce mass-customized products with even a lot size of one. Cost-efficiency, agility, and robustness are, therefore, essential. Processes must have a higher flexibility in case of unexpected events, disturbances, or breakdown, and must even be able to consider changing product/customer requirements during the manufacturing processes.

From the technological point of view, Industry 4.0 has been archived by the consequent integration of concepts of the Internet of Things and Services (Fleisch and Mattern, 2005) and Cyber Physical Systems (Geisberger and Broy, 2012) into production systems. The foundation of the fourth industrial revolution are smart factories with distributed, connected, self-configured, and intelligent systems. Modular production plants and processes allow for a reduction of the overall complexity and increase the level of flexibility and robustness. There is a shift from centralized control to decentralized, autonomous systems which interact in flat hierarchies. Due to this decentralization local problems can often be solved optimally and disturbances are handled where they occur without interfering with other parts of the manufacturing process. Consequently, smart factories provide a higher flexibility and resilience with increased production and quality levels.

In general terms, Industry 4.0 includes a wide range of research and innovation areas, e.g., self-maintaining systems, self-configuration, energy efficiency, human-machine interfaces and interaction, safety, security, 3D-printing, the development of new business models, process analysis and control, law, ergonomics, as well as virtual and augmented reality, to name but a few. In addition, real-time simulation and the modeling of production plants and processes in virtual environments allow for the identification of new optimization potential and also of critical situations at an early stage without disrupting the operating process. This results in a fusion of the real and the virtual world.

1.1 The Importance of Logistics and new Challenges

It is the goal of enterprises to optimize the complete product life-cycle. Therefore, logistics is essential to connect independent autonomous smart factories and to synchronize the production processes, the material flow, and the information flow even beyond company boundaries in the whole supply chain. This includes the transport of raw materials, components, and data between companies as well as the delivery of final products and up-to-date information to the end customers. Thus, logistics is one of the most important parts of Industry 4.0 solutions in order to achieve an inherent optimization and coordination in flexible supply chains (Ten Hompel and Henke, 2014).

The radical change of the economic and logistics structure has already started. The production of bulk goods transported in large quantities by bulk cargo carriers with long-term planning has been decreasing, while the amount of customized high-end products is on the up. Thus, there is a much higher amount of small-size shipments with individual qualities, which must be delivered within a guaranteed time window and probably in a few hours. Consequently, the demand for reliable just-in-time and just-in-sequence transport as well as on courier and express services is increasing while storage capacities and capital commitment are reduced.

This has significant effects on logistics companies whose processes are optimized for cost-efficient transport in permanent structures. They must increase their flexibility to handle the highly volatile order situation and change their processes for application in dynamic environments in which relations and elements of the system continuously change (Lee, 2002). For instance, it is essential to react in real-time to changing traffic conditions, delays at incoming goods departments, and to other dynamically occurring events such as the customer refusing to accept a delivery which then results in a reduced capacity of the vehicle in its ongoing tour. Thus, real-time performance is a fundamental requirement. It implies that planning and scheduling must include the continuous processing of information about the status of the environment and must react to disruptions and unexpected events to ensure the integrity of the process. The efficient computation of an adequate and suitable solution which is determined in real time is preferable to an optimal solution which could be computed too late (Ten Hompel, 2010, pp. 289-290).

Beside the new demands arising from the change of production and manufacturing processes, other *global megatrends* further increase the challenges in logistics (Delfmann and Jaekel, 2012, pp. 6–10). Changing demographics and urbanization lead to an overall grow of commodity flows and also of home-deliveries which are concentrated in urban districts and results in more logistics activities. However, the underlying infrastructures have already reached their limitations, especially in megacities. At the same time, urbanization and future city factories lead to more legal regulations and additional constraints on the usage of the traffic infrastructure. In order to handle the rising transport volume, logistics service providers must increase the capacity utilization significantly and completely avoid empty runs. They must increase their volume, establish new cooperation, and develop innovative concepts, particularly for last-mile logistics and urban deliveries. For example, transport providers must extend their fleet by more heterogeneous and (as yet) unconventional modes of transport such as e-bikes and trams which fit the specific order situation and circumstances. Beside these issues, the economical use of limited natural resources as well as sustainable transportation is essential to reducing the overall costs.

Figure 1.1 illustrates growing customer requirements and their relevance in e-commerce. The fast and cheap delivery of products, increased transparency by track and tracing, the reliability and trust-worthiness of the transport provider, as well as the flexibility to receive the goods at a certain time or time window – these are factors of growing significance. Figure 1.2 also indicates the customer demand for more flexibility of transport service providers in the B2C market. Exactly 70% of people interviewed would like to receive their good at a fixed point in time. In order to reduce waiting times, customers

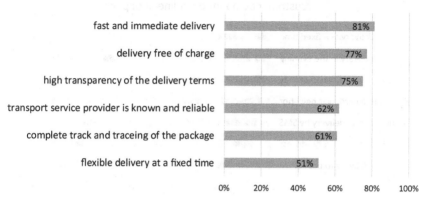

Figure 1.1: Customer requirements and their significance in the consumer's decision to purchase an article by online-shopping of 583 persons interviewed in Germany (Deutsche Post AG, 2012, p. 65).

insist on notification of the exact delivery time in case of delays, e.g., via SMS. Even a short-term change of address must be possible to afford greater freedom of movement and to avoid waiting at a certain location which was defined days ago. The transport company has to adjust its scheduling to the customer's timetable and not vice versa.

In conclusion, the increasing demand on more efficient, high-quality logistics services which consider all domain and customer-dependent constraints as well as higher dynamics significantly aggravates the complexity of logistic planning and control processes.

1.2 Scope of the Thesis and Research Questions

It is the goal of this research to optimize the planning and control processes in order to satisfy rising requirements in transport logistics as shown above. In dynamic supply networks, the interdependencies

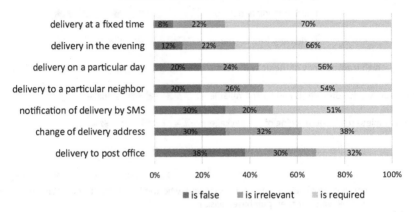

Figure 1.2: Customer demand for more flexibility of the transport service provider in online-purchase of 583 persons interviewed
in Germany (Deutsche Post AG, 2012, p. 63).

and relations of objects change continuously. The challenge is to
consider these interdependencies of numerous system components,
products, and their continuously changing environment to successfully
process, synchronize, coordinate, and optimize the high amount of
decentralized information and material flows even beyond the boundaries of enterprises in the value creation networks. In decentralized
and autonomous systems, this coordination is of growing significance,
because smart objects and products need to plan and control their
way within the production system as well as in the entire supply
chain. Consequently, not only information between enterprises and
customers has to be exchanged, but also information about single
products which directly communicate with transport service providers
and production facilities.

This thesis investigates how the planning, control, and monitoring
of processes can be delegated to the digital representatives of objects

themselves by applying multiagent system technology. It focuses on the following four major research questions:

1. How can autonomous control be implemented into current logistics processes so as to achieve smart transport logistics with low investment?

2. Which communication and negotiation mechanisms allow for multiagent-based autonomous control in transport logistics?

3. How must the decision-making processes of autonomously acting agents be designed to satisfy the requirements of transport service providers and customers in real-time?

4. How large is the optimization potential of industrial partners as a result of implementing multiagent-based autonomous control in their current transport processes?

The main contribution of this research is to develop a general inherent multiagent-based approach that allows for the consideration of domain-dependent requirements to establish smart transport logistics particularly in dynamic real-world environments. Therefore, online performance, stable communication and negotiation mechanisms, and efficient decision-making algorithms are essential. The approach is evaluated on established benchmarks and provides operative examples for two of the most complex and challenging domains in logistics, namely groupage traffic and courier, express, and parcel services (CEP services). In both domains a high amount of decentralized data and information has to be considered, updated, and processed continuously even after the start of transportation. In addition, the heterogeneity of logistic objects and logistic service providers requires the consideration of additional information and constraints of individual entities. The thesis concludes with the results of the transfer project "Autonomous Groupage Traffic" which was funded by the German Research Foundation (DFG) within the Collaborative Research Center 637 (SFB 637). Furthermore, the research was funded by the DFG within the transfer project "Autonomous Courier- and Express Services".

1.2.1 Structure of the Thesis

The thesis is structured as follows. Chapter 2 formally introduces general planning and scheduling problems such as the Traveling Salesman Problem (TSP) and the Vehicle Routing Problem (VRP) in dynamic environments and with various constraints. It discusses classic as well as state-of-the art approaches, mostly centralized, to solve the problems described.

Chapter 3 introduces multiagent systems as well as autonomous logistic processes. It turns out that both are promising to solve complex and dynamic transport problems and show a high reactive and proactive behavior which induces increasing robustness and flexibility. The chapter summarizes the state-of-the-art of multiagent-based approaches in transport logistics and describes the limitations of their applications in Industry 4.0 processes and in real-world operations.

In order to overcome the identified weaknesses of current methods, Chapter 4 presents the dispAgent approach which was developed to meet the increasing requirements on complexity, dynamics, and customization. Firstly, the chapter provides a general overview of its components. Next, it presents stable communication and negotiation protocols for the synchronization in highly parallelized negotiations. In order to satisfy the real-time requirements, efficient routing algorithms were developed which enable the agents to make optimal and also near-to-optimal decisions within a short time (depending on the complexity of the problem). Both algorithms consider domain-dependent constraints and requirements directly within the optimization process. While the first algorithm adapts a classical *branch-and-bound* approach, the second is one of the first to apply the concept of nested Monte Carlo search with policy adaptation to solve routing problems near-to-optimal. The chapter also provides extensive evaluations of the decision-making algorithms by using established benchmarks. Moreover, the chapter investigates the impact of shortest-path searches and multiagent modeling on the overall performance of autonomous processes in transport logistics.

Chapter 5 investigates the influence of parameter variations in different settings and evaluates the system's performance on two established benchmarks for the VRP, namely the Solomon benchmark (Solomon, 1987) and the benchmark of Homberger and Gehring (Homberger and Gehring, 2005).

Next, Chapter 6 provides a case study in groupage traffic which was performed with the Bremen office of Hellmann Worldwide Logistics, one of the biggest forwarding agencies in the world. In contrast to full truckload traffic, in groupage traffic the complexity of process planning is even increased by a highly volatile order situation and changing individual qualities of shipments such as weight, amount, priority, and delivery/pickup time windows. Handling the complexity in real life is aggravated by the high degree of dynamics resulting from unexpected events. The exact amount and properties of shipments are not known in advance. Actual capacities are only revealed and exactly determined during the handling process. Furthermore, undelivered loads in pre-carriage decrease truck capacities in onward carriage. To react to changing traffic conditions and delays at incoming goods departments, it is essential to adapt tours and timetables with regard to actual capacities. Therefore, dynamic scenarios on real-world traffic infrastructures have been simulated by applying multiagent-based simulation. The simulated orders as well as dynamically occurring events are reconstructed from real-world operations. The results reveal that multiagent-based autonomous control meets the sophisticated requirements of groupage traffic.

Due to the fourth industrial revolution and rising demand on immediate delivery, same-day delivery is becoming increasingly important. Therefore, Chapter 7 evaluates the dispAgent approach in a typical application: planning and control of courier, express, and parcel services. This domain in particular, provides one of the most challenging environments because of highly restricted time windows, the demand on immediate transport, and the heterogeneity of goods and vehicles. For the evaluation, real-world data from several days was provided by the industrial partner tiramizoo GmbH. The unmodified data was taken from several service areas and contains all service requests of a complete schedule, including the relevant performance measures of transport tours. The results of the multiagent-based software systems are compared to those of a standard dispatching software product implementing a large neighborhood search. The evaluation shows that the dispAgent approach is competitive and even outperforms the professional standard dispatching product.

Finally, Chapter 8 summarizes and concludes the research. It particularly discusses how smart logistics and Industry 4.0 solutions can profit from the multiagent-based autonomous approach before returning to the initial research questions. The chapter finishes by providing new research perspectives resulting from the present findings which may further improve transport logistics processes.

1.2.2 List of Publications

Content and (preliminary) parts of this thesis as well as more detailed information on related topics were already published in the following book chapters:

- Gath, Herzog, and Edelkamp (2016). Autonomous, Adaptive, and Self-Organized Multiagent Systems for the Optimization of Decentralized Industrial Processes. In J. Kolodziej, L. Correia, and J. M. Molina (Eds.), *Intelligent Agents in Data-Intensive Computing*, Volume 14 of *Studies in Big Data*, pp. 71–98. Springer International Publishing Switzerland.

- Gath, Herzog, and Vaske (2015a). Concurrent and Distributed Shortest-Path Searches in Multiagent-based Transport Systems. Volume 9420 of *Transactions on Computational Collective Intelligence*, pp. 140–157. Springer International Publishing.

- Edelkamp, Gath, Greulich, Humann, and Warden (2014). PlaSMA multiagent simulation. Last-mile connectivity Bangalore. In O. Herzog (Ed.), *German Indian Partnership for IT Systems*, pp. 129–185. München/Berlin: acatech.

article:

- Gath, Edelkamp, and Herzog (2013a). Agent-based dispatching enables autonomous groupage traffic. *Journal of Artificial Intelligence and Soft Computing Research 3* (1), 27–40.

and conference proceedings:

- Gath, Herzog, and Vaske (2016). The Impact of Shortest Path Searches to Autonomous Transport Processes. In H. Kotzab, J. Pannek, and K.-D. Thoben (Eds.), *Dynamics in Logistics – Proceedings of the Fourth International Conference on Dynamics in Logistics* (LDIC), Volume 1, pp. 79–90. Springer International Publishing.

- Gath, Herzog, and Vaske (2015b). Parallel shortest-path searches in multiagent-based simulations with plasma. In S. Loiseau, J. Filipe, B. Dval, and J. van den Herik (Eds.), *Proceedings of the Seventh International Conference on Agents and Artificial Intelligence* (ICAART), Volume 1, pp. 15–21. SciTePress.

- Edelkamp, Gath, Greulich, Humann, Herzog, and Lawo (2015). Monte-Carlo Tree Search for Production and Logistics. In U. Clausen, H. Friedrich, C. Thaller, and C. Geiger (Eds.), *Proceedings of the Second Interdisciplinary Conference on Production, Logistics and Traffic (ICPLT)*, pp. 427–437. Springer International Publishing Switzerland.

- Gath, Herzog, and Edelkamp (2014b). Autonomous and flexible multiagent systems enhance transport logistics. In *Proceedings of the Eleventh International Conference Expo on Emerging Technologies for a Smarter World* (CEWIT), pp. 1–6.

- Edelkamp and Gath (2014). Solving Single-Vehicle Pickup-and-Delivery Problems with Time Windows and Capacity Constraints using Nested Monte-Carlo Search. In B. Duval, J. van den Herik, S. Loiseau, and J. Filipe (Eds.), *Proceedings of the Sixth International*

Conference on Agents and Artificial Intelligence (ICAART), Volume 1, pp. 22–33. SciTePress.[1]

- Edelkamp, Gath, and Rohde (2014). Monte-carlo tree search for 3d packing with object orientation. In C. Lutz and M. Thielscher (Eds.), *KI 2014: Advances in Artificial Intelligence*, Volume 8736 of *Lecture Notes in Computer Science*, pp. 285–296. Springer International Publishing.

- Gath, Herzog, and Edelkamp (2013). Agent-based planning and control for groupage traffic. In *Proceedings of the Tenth International Conference and Expo on Emerging Technologies for a Smarter World* (CEWIT), pp. 1–7.

- Edelkamp, Gath, Cazenave, and Teytaud (2013). Algorithm and knowledge engineering for the TSPTW problem. In *IEEE Symposium on Computational Intelligence in Scheduling* (SCIS), pp. 44–51.

- Gath, Edelkamp, and Herzog (2013b). Agent-based dispatching in groupage traffic. In *Proceedings of the IEEE Workshop on Computational Intelligence in Production and Logistics Systems* (CIPLS), pp. 54–60.

- Edelkamp and Gath (2013). Optimal Decision Making in Agent-Based Autonomous Groupage Traffic. In J. Filipe and A. L. N. Fred 176 Bibliography (Eds.), *Proceedings of the Fifth International Conference on Agents and Artificial Intelligence* (ICAART), Volume 1, pp. 248–254. SciTePress.

- Greulich, Edelkamp, and Gath (2013). Agent-Based Multimodal Transport Planning in Dynamic Environments. In I. J. Timm and M. Thimm (Eds.), *KI 2013: Advances in Artificial Intelligence*,

---

[1]For this article and its presentation Max Gath received the *Best Student Paper Award* at the International Conference for Agents and Artificial Intelligence in 2014.

Volume 8077 of *Lecture Notes in Computer Science*, pp. 74–85. Springer Berlin Heidelberg.

- Greulich, Edelkamp, Gath, Warden, Humann, Herzog, and Sitharam (2013). Enhanced shortest path computation for multiagent-based intermodal transport planning in dynamic environments. In J. Filipe and A. L. N. Fred (Eds.), *Proceedings of the Fifth International Conference on Agents and Artificial Intelligence* (ICAART), Volume 2, pp. 324–329. SciTePress.

- Lewandowski, Gath, Werthmann, and Lawo (2013). Agentbased Control for Material Handling Systems in In-House Logistics – Towards Cyber-Physical Systems in In-House-Logistics Utilizing Real Size. In *Proceedings of the European Conference on Smart Objects, Systems and Technologies* (SmartSysTech), pp. 1-5.

- Gath, Wagner, and Herzog (2012). Autonomous logistic processes of bike courier services using multiagent-based simulation. In M. Affenzeller, A. Bruzzone, F. D. Felice, D. D. R. Vilas, C. Frydman, M. Massei, and Y. Merkuryev (Eds.), *Proceedings of the Eleventh International Conference on Modeling and Applied Simulation*, pp. 134–142.

In addition, these articles focus on the applicability of the developed multiagent-based approach in real-world scenarios:

- Gath and Herzog (2015). Intelligent Logistics 2.0. *German research magazine of the Deutsche Forschungsgemeinschaft 1*, 26–29.

- Gath, Herzog, and Edelkamp (2014a). Agenten für eine optimierte Logistik. *RFID im Blick Sonderausgabe Industrie 4.0 und Logistik 4.0 aus Bremen*, 36–37.

2 Dispatching Problems in Transport Logistics

It is the goal of optimization to adopt a system behavior in order to maximize or minimize an objective function. In economic systems, this objective function refers to revenue maximization. Logistics is a classical area of application for optimization, because it includes multiple complex optimization problems. In transport logistics, these complex problems often refer to tour planning and routing problems. The goal of transport is, to pick up goods at their origin and deliver them to their destination to overcome special distances in order to fulfill the six objectives of logistics: to deliver the right object (1) at the right time (2) to the right place (3) in the right quantity (4) and quality (5) at the right price (6) (Jünemann, 1989, p. 18). Consequently, most logistics problems require multicriteria optimization taking into account several constraints.

This chapter introduces tour planning and routing problems in transport logistics and focuses on the most relevant problem variations and established approaches to solve them. Therefore, the problems are categorized in three groups. Firstly, Section 2.1 presents the well-known Traveling Salesman Problem (TSP) with various constraints which has the objective of determining the shortest route for a single vehicle which includes all stops and fulfills additional constraints. Section 2.2 describes multivehicle variants of the Vehicle Routing Problem (VRP). Its main difference is that not only the shortest tour, but also the optimal allocation of goods to vehicles must be determined. Section 2.3 presents the dynamic variant of the VRP. Since there are numerous domain-dependent variations of each problem and at least as many approaches to solve them, comprehensive overviews are referenced.

Formal descriptions of mathematical programs are provided as long as they are relevant in subsequent chapters. Section 2.4 concludes the chapter and discusses the limitations of applying classic approaches in dynamic real-world and Industry 4.0 applications.

2.1 Traveling Salesman Problems

The most famous problem in logistics is probably the Traveling Salesman Problem (Flood, 1956), which is NP-complete (Garey, Graham, and Johnson, 1976). The classical TSP can be defined as follows.

Definition 2.1 (Traveling Salesman Problem). *Let S denote a set of stops, which must be visited. Given the costs $c_{i,j}^v$ for traveling from $i \in S$ to $j \in S$ and choosing indicator variables*

$$x_{i,j} = \begin{cases} 1, & \text{if } (i,j) \text{ is part of the vehicle's tour} \\ 0, & \text{otherwise} \end{cases} \tag{2.1}$$

the objective function of the classical TSP is

$$\min \sum_{j \in S} \sum_{i \in S} c_{i,j} \cdot x_{i,j} \tag{2.2}$$

subject to

$$x_{i,j} = \{0,1\} \text{ for all } i, j \in S \tag{2.3}$$

$$\sum_{i \in S} x_{i,j} = 1 \text{ for all } j \in S \tag{2.4}$$

$$\sum_{j \in S} x_{i,j} = 1 \text{ for all } i \in S \tag{2.5}$$

$$\sum_{j \in S} \sum_{i \in S} x_{i,j} \leq |Y| - 1 \text{ for all } Y \subseteq S. \tag{2.6}$$

The objective function defined by Eq. 2.2 minimizes the overall cost. Eq. 2.4 and Eq. 2.5 ensure that each service request has to be satisfied exactly once. In addition, Eq. 2.6 is a sub-tour elimination constraint which guarantees that all stops are visited in a single and connected tour. In short, the goal is to find the shortest tour of a vehicle, which ensures that all service requests are satisfied.

In most real-world scenarios, time window constraints as well as time consumption at the warehouse/customer have to be considered. The Traveling Salesman Problem with Time Windows (TSPTW) which also involves handling times can be defined as follows.

Definition 2.2 (TSP with Time Windows and Handling Times). *If l_s denotes the latest pickup/delivery time at stop $s \in S$, t_s the time consumption of the loading or unloading process, r_s the release time at s, and $time_{i,j}$ vehicle's time for driving from i to j, then*

$$x_{i,j} = 1 \Rightarrow l_j \geq r_i + t_j + time_{i,j} \qquad (2.7)$$

has to be fulfilled.

Let $DEP \subseteq S$ denote a set of depots. In some TSPTW it is relevant that the vehicle starts and ends at a depot $d \in DEP$. This is ensured by the following equations.

$$r_d < \min_j r_{j \in S \setminus DEP} \qquad (2.8)$$

and

$$l_d > \max_j r_{j \in S \setminus DEP}. \qquad (2.9)$$

In some TSPs capacity constraints have to be considered.

Definition 2.3 (TSP with Capacity Constraints). *Let CC_s denote the current capacity of the vehicle at stop $s \in S$ and M the maximum capacity of the vehicle, then*

$$CC_s \leq M \text{ for all } s \in S \qquad (2.10)$$

has to be fulfilled.

Of course, there are problems which include time windows, handling times (Eq. 2.7), and capacity constraints (Eq. 2.10) at the same time. In addition, there are TSPs with pickup and delivery requests at the stops. If the transport is offered between a central depot and customers, it is not relevant which kind of service is offered at a stop as long as the maximum allowed capacity of the vehicle is not exceeded at any stop (Eq. 2.10). However, if the problem includes transports between customers in direct tours, the vehicle has to ensure that it visits the pickup stop before the delivery stop. This kind of Traveling Salesman Problem with Pickup and Deliveries (TSPPD) can be defined as follows.

Definition 2.4 (TSP with Pickups and Deliveries). *Stops are a super set of a subset of pickup stops $P \subset S \setminus (D \cup DEP)$ and a super set of a subset of delivery stops $D \subset S \setminus (P \cup DEP)$. Moreover, O denotes a set of orders. An order $o \in O$ contains exactly one pickup stop p_o and one delivery stop d_o. Therefore, it must be ensured that the vehicle visits the pickup stop before the delivery stop by*

$$(x_{i,p_o} = 1 \wedge x_{i,d_o} = 1) \Rightarrow l_{p_o} < r_{d_o}. \tag{2.11}$$

Definitions 2.1 to 2.4 describe the TSP with frequently considered constraints. Nevertheless, there could be numerous additional constraints, especially in real-world problems. For instance, Jaillet (1988) solves the TSP with probabilistic customer demands and Malandraki and Dial (1996) present a dynamic programming heuristic for solving the time-dependent TSP.

The TSP and its variations are among the most investigated problems in mathematics, Operations Research (OR), and computer science, because the TSP is NP-hard, easy to describe, and hard or even impossible to solve optimally. There are numerous approaches to solve the TSP with various constraints such as genetic algorithms (Grefenstette, Gopal, Rosmaita, and Van Gucht, 1985; Potvin, 1996; Snyder and Daskin, 2006), simulated annealing (Aarts, Korst, and van Laarhoven, 1988; Geng, Chen, Yang, Shi, and Zhao, 2011), tabu-search (Fiechter, 1994; Gendreau, Guertin, Potvin, and Taillard, 1999;

Gendreau, Laporte, and Semet, 1998), ant colony systems (Dorigo and Gambardella, 1997), particle swarm algorithms (Shi, Liang, Lee, Lu, and Wang, 2007), neural networks approaches (Potvin, 1993; Créput and Koukam, 2009), k-opt improvement heuristics (Lin and Kernighan, 1973; Helsgaun, 2009), as well as branch-and-bound algorithms (Padberg and Rinaldi, 1991; Fischetti, Salazar González, and Toth, 1997; Hernández-Pérez and Salazar-González, 2004; Cordeau, Iori, Laporte, and Salazar González, 2010) with several heuristics for upper and lower bounds (Karp and Steele, 1985; Miller and Pekny, 1991; Johnson and McGeoch, 2007), to name but a few. A comprehensive review of the TSP is provided by, e.g., Applegate, Bixby, Chvatal, and Cook (2006) and Cook (2012).

2.2 Vehicle Routing Problems

The Vehicle Routing Problem (VRP) (Dantzig and Ramser, 1959) is a generalization of the TSP which includes multiple vehicles. While the objective of the TSP is to find the shortest route for a single vehicle, the objective of the VRP is to find the optimal allocation of goods to vehicles as well as the shortest route for each vehicle, both of which minimize the total costs. The VRP is an NP-hard problem (Lenstra and Kan, 1981). It is hard to solve even small problems with 25-50 stops optimally (Azi, Gendreau, and Potvin, 2010).

Similar to Definition 2.1 the VRP can be defined as follows.

Definition 2.5 (Vehicle Routing Problem). *Let V denote a set of vehicles. Given the costs $c_{i,j}^v$ for a vehicle $v \in V$ for traveling from $i \in S$ to $j \in S$ and choosing indicator variables*

$$x_{i,j}^v = \begin{cases} 1, & \text{if } (i,j) \text{ is part of vehicle } v\text{'s tour} \\ 0, & \text{otherwise} \end{cases} \tag{2.12}$$

the highest priority objective function of a VRP is to minimize the number of used vehicles by

$$\min \sum_{v \in V} 1 \tag{2.13}$$

*and second highest priority is to minimize the cost of the vehicles,
which commonly depends on the distances driven or travel times of
the vehicle, by*

$$\min \sum_{v \in V} \sum_{j \in S} \sum_{i \in S} c_{i,j}^v \cdot x_{i,j}^v \qquad (2.14)$$

subject to

$$\sum_{v \in V} \sum_{i \in S} x_{i,j}^v = 1 \ for \ all \ j \in S \qquad (2.15)$$

$$\sum_{v \in V} \sum_{j \in S} x_{i,j}^v = 1 \ for \ all \ i \in S \qquad (2.16)$$

$$\sum_{v \in V} x_{i,j}^v = \{0,1\} \ \ for \ all \ i,j \in S \qquad (2.17)$$

$$\sum_{v \in V} \sum_{j \in S} \sum_{i \in S} x_{i,j}^v \leq |Y| - 1 \ for \ all \ Y \subseteq S. \qquad (2.18)$$

Depending on additional constraints, Eq. 2.7 - Eq. 2.11 must be
satisfied for each vehicle $v \in V$ as well.

If the VRP also includes direct tours between customers without any
handling operations at a central depot, it is known as the Pickup and
Delivery Problem (PDP). While the classical PDP considers the transport of all kinds of goods, the so-called Dial-a-Ride Problem(DARP)
(Cordeau and Laporte, 2007) or Handicapped Persons Transportation
Problem (Toth and Vigo, 1997) deals with passenger transport and
additional objective functions such as minimizing the transport times
of passengers.

In unpaired PDPs transported goods are homogeneous and ex-
changeable. Thus, any item can be delivered to any customer. In
paired PDPs every item has a specific sender and recipient. Consequently, the pickup and delivery requests of an order o have to be
served by the same vehicle v. This is guaranteed by

$$\sum_{i \in S} x_{i,p_o}^v - \sum_{i \in S} x_{i,d_o}^v = 0 \ \text{for all} \ i \subseteq S \ \text{and} \ v \in V. \qquad (2.19)$$

Problems which contain multiple depots (Renaud, Laporte, and Boctor, 1996) can be considered as a specialization of a PDP described above. In general, the variations of possible constraints of VRPs are similarly diversified as those of the TSP. For instance, there are problems with soft time windows (Balakrishnan, 1993; Taillard, Badeau, Gendreau, Guertin, and Potvin, 1997) and problems with probabilistic customer demands (Gendreau, Laporte, and Séguin, 1996). Especially in real-world problems, the complexity is increased by domain-dependent constraints and requirements. For instance, having a heterogeneous vehicle fleet requires the consideration of the individual properties of the vehicles in the tour planning process. They vary in their capacities. They differ in their costs per kilometer. They have individual working times. Some freight carriers are paid by commission fee while others receive a fixed rate per day. Moreover, loads are generally more individualized than is assumed in scientific investigations. The goods vary not only in their pickup or delivery time window and weight but also in their value, priority, and volume. In addition toll and dangerous goods restrictions as well as customs guidelines on security must be satisfied. An abstract illustration of a solution for a static VRP is shown in Figure 2.1.

Theorem 2.1 (Complexity of the VRP). *Let k denote the number of vehicles minus 1 ($k = |V| - 1$) and n denote the amount of stops which have to be visited ($n = |S|$). Then, the number of possible tours for the VRP is at most $\frac{(n+k)!}{k!} = O(n!(n + k)^k)$.*

Proof. In order to illustrate the complexity and the amount of possible solutions of a general VRP, imagine a bookshelf on which you have to stack a number of books. In addition, the bookshelf can be divided into several sections by separators. In this example, the books refer to the amount of stops (denoted as n) which have to be visited and each section refers to a vehicle. Thus, there are $|V|$ sections and $|V| - 1$ separators (denoted as k) as shown in Figure 2.2. If there is no separator, there is only a single vehicle and there are $n!$ combinations to stack the books. If there is a single separator, there are $n!\binom{n+1}{1}$ combinations (all possible combinations of books which result from

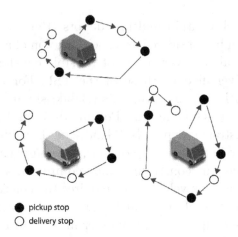

Figure 2.1: An abstract visualization of a solution for a static VRP with pickups and deliveries.

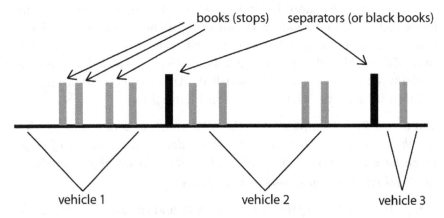

Figure 2.2: The figure shows an illustration of a VRP with 3 vehicles and 9 stops applied to the *bookshelf* example. There are $1,320$ possibilities for assigning the books to the sections divided by the separators. In other words, there are $\binom{9+3}{3}$ combinations to put the separators for each of the 3! permutations of the books positions.

Table 2.1: The table shows the number of possible combinations for very small-size general VRPs. Note, that in problems with orders having a pickup and delivery, e.g., 10 stops are required to transport 5 orders only.

vehicles stops	1	2	3	4	5
1	1	2	3	4	5
2	2	6	12	20	30
3	6	24	60	120	210
4	24	120	360	840	1,680
5	120	720	2,520	6,720	15,120
6	720	5,040	20,160	60,480	151,200
7	5,040	40,320	181,440	604,800	1,663,200
8	40,320	362,880	1,814,400	6,652,800	19,958,400
9	362,880	3,628,800	19,958,400	79,833,600	$\approx 2.59 \cdot 10^8$
10	3,628,800	39,916,800	$\approx 2.39 \cdot 10^8$	$\approx 1.03 \cdot 10^9$	$\approx 3.63 \cdot 10^9$
11	39,916,800	$\approx 4.79 \cdot 10^8$	$\approx 3.11 \cdot 10^9$	$\approx 1.45 \cdot 10^{10}$	$\approx 5.44 \cdot 10^{10}$
12	$\approx 4.79 \cdot 10^8$	$\approx 6.22 \cdot 10^9$	$\approx 4.35 \cdot 10^{10}$	$\approx 2.17 \cdot 10^{11}$	$\approx 8.71 \cdot 10^{11}$
13	$\approx 6.22 \cdot 10^9$	$\approx 8.71 \cdot 10^{11}$	$\approx 6.53 \cdot 10^{11}$	$\approx 3,48 \cdot 10^{12}$	$\approx 1.48 \cdot 10^{13}$
14	$\approx 8.71 \cdot 10^{11}$	$\approx 1.30 \cdot 10^{12}$	$\approx 1.04 \cdot 10^{13}$	$\approx 5.92 \cdot 10^{13}$	$\approx 2.66 \cdot 10^{14}$

different locations of the separators). If there are two separators, there are $n!\binom{n+2}{2}$ combinations. Consequently, with k separators there are $n!\binom{n+k}{k}$ combinations. The value $n!\binom{n+k}{k}$ can be transformed by

$$n!\binom{n+k}{k} = n!\frac{(n+k)!}{n!k!} = \frac{(n+k)!}{k!} \qquad (2.20)$$

due to the factorial equation of binomial coefficients $\binom{n'}{k} = \frac{n'!}{k!(n'-k)!}$. To follow the division of $k!$, the separators can be shown as black books which are interchangeable (in contrast to the other books which are unique). Due to $\binom{n'}{k} \leq n'^k$, the effort is $O(n!(n+k)^k)$. Table 2.1 depicts the numbers of possible assignments for very small-size VRPs computed by Eq. 2.20. □

Optimal approaches to solve the VRP often apply *branch-and-bound* techniques, which are limited to solving small problems due to the high computational complexity. For instance, Ropke, Cordeau, and

Laporte (2007) present an exact solution method for PDPs containing 96 orders. The solver of Azi et al. (2010) computes optimal solutions for VRPs with 25-50 stops. Over the few last decades, numerous efficient heuristics and meta-heuristics have been developed for solving large VRPs and PDPs such as tabu-searches (Garcia, Potvin, and Rousseau, 1994; Rochat and Taillard, 1995; Taillard et al., 1997; Cordeau and Laporte, 2003), genetic algorithms (Potvin and Bengio, 1996; Tan, Lee, Ou, and Lee, 2001; Berger and Barkaoui, 2004; Bräysy, Dullaert, and Gendreau, 2004; Pankratz, 2005), large neighborhood searches (Shaw, 1998; Ribeiro and Laporte, 2012), simulated annealing (Bent and Hentenryck, 2006), and ant systems (Gambardella, Éric Taillard, and Agazzi, 1999; Barán and Schaerer, 2003; Gajpal and Abad, 2009), to name but a few. Comprehensive and recommendable overviews of solution methods for solving multiple variations of VRPs are provided by, e.g., Bräysy and Gendreau (2005a), Bräysy and Gendreau (2005b), Golden, Raghavan, and Wasil (2008), and Parragh, Doerner, and Hartl (2008a,b).

An early publication by Clarke and Wright (1964) applies the so-called *savings heuristic*. The algorithm assigns each order to a vehicle which must only transport the respective order. In following iterations, the tours of two vehicles are merged to a single tour as long as all constraints are satisfied. This process is continued until the maximum capacity of a vehicle is reached. A frequently applied and established heuristic for the construction of tours is the *sequential insertion heuristic I1* developed by Solomon (1987) which is based on the savings heuristic. Solomon's algorithm or a variation of it is applied by several of meta-heuristics referenced above and also by numerous of the multiagent-based approaches which are described in Section 3.4. Solomon's algorithm initially creates a new tour containing only a high constrained order. Using a constrained order (instead of a less constrained one) increases the probability of inserting a higher number of less constrained orders in the following steps. Next, for each not yet inserted order the minimum cost and optimal position in the constructed tour is computed. The order with the least cost is inserted at the respective position. This process is continued until the orders' constraints prohibit inserting any additional order. The goal of the sequential insertion heuristic I1 is to minimize the number of vehicles required to transport all orders by satisfying all constraints. In addition, Solomon (1987) presents the *sequential insertion heuristic I2* to minimize the distances driven and the *sequential insertion heuristic I3* to minimize customers' waiting times of customers. Several authors, such as Dullaert and Bräysy (2003), Potvin and Rousseau (1993), and Balakrishnan (1993), adapt the insertion algorithm, e.g., to allow for parallel insertion, to consider soft time windows, or to increase performance.

As the tour's efficiency highly depends on the sequence in which orders are inserted in the tours, the algorithms are neither optimal nor complete. To reduce this dependency, k-Opt improvement techniques (Lin, 1965a) are applied such as the 2-Opt or 3-Opt operators (Solomon and Desrosiers, 1988; Savelsbergh, 1992), the 2-Opt* operator (Potvin and Rousseau, 1995), or the CROSS-Exchange operator (Taillard et al., 1997). For example, the tabu-search approach of Garcia et al. (1994) applies the I1 heuristic of Solomon (1987) to compute an initial solution, which is further optimized by k-opt improvement techniques.

2.3 Dynamic Vehicle Routing Problems

The classical VRP described above is a static and deterministic problem. All orders, vehicles, constraints, and the distance matrix are exactly known in advance and do not change. However, most real-world applications in transport logistics refer to dynamic problems in which information is revealed or updated during operations after an initial solution is computed. Thus, the planning horizon is generally unbounded, and scheduling and planning requires a periodical or continuous reoptimization to react to new or changing information and unexpected events. An example for solving a dynamic routing problem is illustrated in Figure 2.3.

In general, dynamic events can be categorized in three different classes:

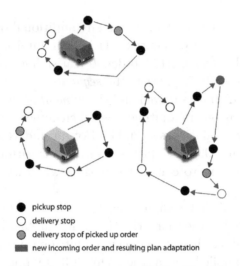

 pickup stop
○ delivery stop
◑ delivery stop of picked up order
▬ new incoming order and resulting plan adaptation

Figure 2.3: A solution for the dynamic VRP with pickup and deliveries.

1. dynamically incoming or outgoing entities, e.g., orders or vehicles;

2. dynamically occurring unexpected events;

3. new or changing information transmitted dynamically.

Since customer demands are not fully known in advance, most investigations only focus on dynamically incoming orders which have to be integrated into existing tours. However, real-world planning and control often need to consider some of the following dynamic events as well:

- The amount, type, and properties of the vehicle fleet change during operations. This is particularly relevant in domains with an order situation that varies daily or seasonally.

- The exact amount of goods which must be picked up at a customer location could be significantly larger than estimated. If the plan is not adapted, loading all pallets might result in exceeding the maximum capacity of the vehicle during the rest of the trip. Also a breakdown or a delay in the manufacturing process could reduce the real demand. Thus, available capacities are wasted if the system is not re-optimized.

- Other properties of the goods may change. For instance, there is increasing customer demand in courier- and express services to change the pickup or delivery address dynamically (Deutsche Post AG, 2012, p. 63). Also time windows should be adapted dynamically to offer more flexible and thus customer-friendly services.

- Delay in an incoming goods department means further delays on the overall tour plan.

- The traveling time may vary, e.g., because of traffic congestions.

While this list includes general aspects only, further domain specific demands have to be considered in real-world operations.

In most dynamic problems, some information is already known in advance, while additional information becomes known during the operation. In order to measure the dynamics of a problem, Lund, Oli, and Rygaard (1996) introduced the *Degree of Dynamism*.

Definition 2.6 (Degree of Dynamism). *Let n_{con} denote to number of dynamically appearing events, e.g., the number of available orders, and let n_{total} denote the total number of events, e.g., the overall amount of orders. The Degree of Dynamism is defined by*

$$dod = \frac{n_{con}}{n_{total}}. \tag{2.21}$$

The higher the *dod* value is, the more dynamic is the system. The *dod* assumes that each dynamic event during operation occurs evenly distributed. However, this depends on the application domain. For instance, 90% of all orders might come in during the rush hour after the operation has started. Therefore, Larsen, Madsen, and Solomon (2002) introduced the *Effective Degree of Dynamism*.

Definition 2.7 (Effective Degree of Dynamism). *The planning horizon is limited by T. Let $t \in \mathbb{R}$ with $0 \leq t \leq T$ define the point of time*

an event occurs during operation. The Effective Degree of Dynamism
is defined by

$$edod = \frac{\sum\limits_{i=1}^{n_{con}} \left(\frac{t_i}{T}\right)}{n_{total}}. \tag{2.22}$$

Thus, the edod value can also take into account rush hours and
peak buying times.

Solving dynamic transport problems instead of static problems increases the flexibility and robustness in real-world transport processes
and improves the customer service quality. Moreover, adapting to the
current situation creates more optimization potential than advanced
planning with estimated ranges of variable values. Thus, the overall
efficiency increases and the operational costs decrease. Lack of information, unavailability of data, and uncertainty in the environment
are handled dynamically. However, the computational complexity
is further increased by a high degree of dynamics. Comprehensive
surveys of dynamic VRPs are provided by, e.g., Ghiani, Guerriero,
Laporte, and Musmanno (2003), Berbeglia, Cordeau, and Laporte
(2010), and Pillac, Gendreau, Guret, and Medaglia (2013).

2.3.1 Static Approaches for Dynamic Problems

One of the first approach for solving dynamic VRPs was designed
by Psaraftis (1980). His approach make use of a static and optimal
dynamic programming solver which performs immediate replanning
whenever a dynamic event appears. Thus, the dynamic problem is
split up into multiple static problems. Each static sub-problem is
solved optimally and all available information is taken into account.
However, the optimal dynamic programming solver has a computational complexity of $O(n^2 3^n)$. Consequently, the complexity of VRPs
limits the approach to solving only very small problems.

Motivated by this approach, other authors such as Savelsbergh and
Sol (1998), Yang, Jaillet, and Mahmassani (2004), Montemanni, Gambardella, Rizzoli, and Donati (2005), and Chen and Xu (2006) followed
the general idea of splitting up the dynamic problem into several time

episodes, time slices, or time epochs. Next, a static solver computes solutions for each of these sub-problems independently. However, also the application of meta-heuristics and non-optimal approaches has significant drawbacks. Even marginally relevant events cause a complete recomputation of a new NP-hard problem from scratch. Especially with larger problems, the resulting response times are unacceptable for dispatchers and operators in real-world applications. Another disadvantage is that even small improvements could result in complete reallocations of shipments to vehicles, because each problem is solved independently. This complicates the handling process and could confuse the operators. To avoid this problem Montemanni et al. (2005) facilitate the reassignment of orders by applying a *Least Commitment Strategy* which assigns orders at the latest possible time. This is when a vehicle is idle or a later transport would violate time constraints. However, applying a *Least Commitment Strategy* decreases the service quality through late deliveries and is inefficient in scenarios where shipments can be consolidated in economical tours. Instead, this strategy could successfully solve the dynamic Stacker Cane Problem (SCP) (Frederickson, Hecht, and Kim, 1976). The dynamic SCP is a specialization of the dynamic PDP in which vehicles have a limited capacity of exactly one. However, it is not possible to react to other dynamically appearing events on the way to the next stop.

2.3.2 Continuous Reoptimization

Another approach to solve dynamic problems is to first compute a solution for the static problem and continuously update this solution whenever a dynamic event appears. Gendreau et al. (1999) developed a tabu-search algorithm which uses a continuous reoptimization of tours to solve the dynamic VRP. They adapted the method of Rochat and Taillard (1995) who implemented the so-called *adaptive memory*, a temporary storage space for best solutions already found. Similar to genetic algorithms, a fitness function evaluates the efficiency of each tour. The tours of the adaptive memory are adapted or combined and the resulting new route is added to the adaptive memory. As the size of the storage is bounded, only best solutions remain saved. Gendreau et al. (1999) handle the dynamics by firstly solving a static problem. If a dynamic event appears, all tours which are saved in the adaptive memory are updated and the optimization process is continued. However, a *least commitment strategy* is applied to assign the orders to the vehicles. Thus, the approach has all the disadvantages of *least commitment strategies* described above. Other authors have implemented similar tabu-search and large neighborhood search algorithms for solving dynamic VRPs such as Kergosien, Lent, Piton, and Billaut (2011), Mitrovic-Minic and Laporte (2004), Ichoua, Gendreau, and Potvin (2006), as well as Gendreau, Guertin, Potvin, and Séguin (2006).

Other authors applied one of the approaches described in Section 2.2 to initially solve the static problem. Next, dynamically incoming orders are integrated into existing tours by insertion heuristics, by k-opt improvement techniques, and by partial modifications of tours like Jih and Hsu (1999), Montané and ao (2006), Hanshar and Ombuki-Berman (2007), Beaudry, Laporte, Melo, and Nickel (2010). As these approaches have an initial population which is optimized by mutation and combination techniques, they are also called genetic algorithms. Unfortunately, in most cases the approaches consider only the integration of new orders or vehicles and neglect other dynamic events.

2.3.3 Anticipation and Stochastic Approaches

Stochastic approaches process historical data to anticipate future events. On the one hand, in static stochastic problems anticipations can be used to find adequate a-priori solutions which are only slightly modified during operations. In general, these adaptations are performed manually. For example, problems can include stochastic customer demands, which are estimated on the basis of a probability distribution, while the exact demand is revealed during operations. On the other hand, predictions and the consideration of future situations and events improve the proactive behavior of the system and reduce the required reactiveness in dynamic stochastic problems. Especially in domains with fix customers and recurring events, companies benefit from anticipations, e.g., of dynamically incoming orders. This increases the overall efficiency of the planning process.

Ichoua et al. (2006) consider probabilistic information about future events in the tour construction process. Therefore, they split up the service area into multiple clusters. In order to optimize the allocation of vehicles to these clusters, they consider not only the current status and information, e.g., about the actual distribution of vehicles and the current order situation, but also the probability of future incoming orders in the decision making process. Thus, the allocation is continuously optimized to shorten the reaction time and to increase the resource utilization to satisfy all service requests in each cluster.

Similarly, other authors apply anticipations in combination with different *waiting strategies* such as Mitrovic-Minic and Laporte (2004), Branke, Middendorf, Noeth, and Dessouky (2005), Ichoua et al. (2006), Thomas (2007), and Pureza and Laporte (2008). If a vehicle has finished the handling operations at a certain stop, it remains idle instead of immediately continuing the tour. The idea is that in anticipation of new incoming orders in nearby districts, the reaction time, the overall required time, and the distances driven can be reduced. Although the authors prove the advantages of *waiting strategies* in their investigated domains, the benefit highly depends on the quality of anticipation and the structure of the problem. In most real-world cases it is unacceptable to wait at specific stops or to increase the idle times of vehicles if it is not guaranteed that new orders will appear. In addition, waiting requires enough time to visit the next stop on time and probably reduces the service quality, because goods are delivered at the latest possible time.

2.4 Summary and Conclusion

This chapter introduced general as well as the most relevant dispatching problems in transport logistics such as TSP, VRP, and dynamic VRP including numerous variations with diversified constraints and requirements. Since route and tour planning is a classic application domain of optimization algorithms, the general problems with standard constraints have intensively been investigated.

Operations Research (OR) and mathematical approaches such as dynamic programming clearly dominated the research in logistics planning and scheduling at the end of the 20th century. However, especially in the last few decades numerous efficient algorithms have been developed for the transportation domain, extending these approaches with artificial intelligence. Therefore, numerous heuristics and meta-heuristics, such as simulated annealing, tabu-search, neural networks, ant systems, and genetic algorithms, extend and combine OR methods to both reduce the search space and to accelerate the search. Moreover, they increase the quality of solutions for large problems. These mainly centralized solvers have been extensively evaluated and also been applied in professional software systems for transport planning and scheduling.

Motivated by technological advances such as the Global Positioning System (GPS), smartphones, Geographical Information Systems (GIS), and cheap communication technologies which cover larger areas, research started to focus on the dynamic VRP around the year 2000. However, most of the approaches developed so far limit their examinations of dynamics to continuously incoming orders. Other dynamic events, such as dynamically changing addresses and order information, delays at incoming goods departments, and breakdowns of vehicles, are often neglected.

In order to handle the dynamics, a general approach presented in Section 2.3.1 is to split up the dynamic problem in a sequence of multiple static problems. However, the main drawback is that this results in long reaction times for dispatchers and freight carriers, due to the high computational complexity of the static problems. Especially in dynamic problems, online and short-term decision-making is essential. In addition, freight carriers are confused if allocations change significantly, which frequently happens if solutions are recomputed from scratch in each time episode. These drawbacks reduce the usability of these systems and preclude their application in real-world operation.

Therefore, Section 2.3.2 discusses approaches which continuously re-optimize a solution in case of dynamically appearing events. Many of the methods presented have a restricted proactive behavior, because they apply a *least commitment strategy*. This *wait and go* strategy impedes the consolidation of goods to build more economical loads, but might be successfully applied for problems which only include full-truckload storage. K-opt improvement techniques in combination with other meta-heuristics, such as simulating annealing or large neighborhood searches, are often applied and result in adequate solutions. Nevertheless, similar to most other centralized approaches, there is a lack of individualization: The solutions fail to be easily adapted and modified by additional domain dependent constraints because of the high complexity of the model and/or of the equations. In general, they even neglect considering a heterogeneous fleet. In addition, most approaches for dynamic problems are evaluated on benchmarks. Thus, they only use Euclidian distances between stops and locations to compute the required distance-matrix as input. They ignore the problem and the computational effort of recurring shortest-path computations. Especially in real-world operation, efficient shortest-path computations are essential (cf. Section 4.5).

Stochastic approaches presented in Section 2.3.3 are promising methods to further extend and improve other solvers. In problems with a limited degree of dynamics which include multiple periodical incoming orders on a regular basis, they can identify patterns of dynamic events. Thus, anticipations are applied to transform the dynamic problem into a stochastic static one, whose solution has merely (and often manually) been adapted during operations.

In conclusion, classical centralized planning and control is limited in dynamic, complex, and customized logistics processes due to the requirements of flexibility and adaptability to changing environmental influences and individual processes. Autonomous logistic processes and multiagent systems overcome these essential drawbacks.

3 Multiagent-Based Transport Planning and Control

As described in Chapter 1, the *Fourth Industrial Revolution* has increased the complexity and dynamics in logistics due to shorter product life-cycles, rising numbers of product variants, and growing numbers of links and dependencies of production processes between companies. This particularly affects the transport processes within production and supply chain networks.

To satisfy customer demands, transport service providers have to increase their service quality while countering the increasing cost pressure in the logistics sector. Consequently, they have to consider more individual properties of customized shipments, transport higher amounts of small-size orders, and guarantee shorter transit times. In addition, the rising traffic density of transport infrastructures and growing demands with regard to sustainable transportation encourage logistics companies to improve their process efficiency.

As a result, the increasing dynamics and complexity of (decentralized) planning and scheduling processes cannot be handled by human operators or by centralized decision-making software. They require an efficient, proactive, and reactive system behavior. Moreover, individualized services, such as services offering a wide range of products (cf. case study in Chapter 6) or courier and express services (cf. case study in Chapter 7), force transport service providers to implement customized and adaptive business processes. However, the dynamics in logistics and individual requirements of diversified application domains are often neglected by automated scheduling approaches.

While centralized planning and control in complex and dynamic processes are increasingly difficult due to the requirements of flexibility and adaptability in multiple changing environments, multiagent systems (MASs) are able to overcome these essential obstacles. They can be applied to solve complex, dynamic, and distributed problems (Müller, 1997) in which agents are a natural metaphor for physical objects and actors (Wooldridge, 2009, pp. 183-184). The advantages of applying multiagent systems are high flexibility, adaptability, scalability, and robustness in decentralized systems, which are achieved by problem decomposition and the proactive, reactive, and adaptive behavior of intelligent agents. Thus, multiagent systems (cf. Section 3.1) and multiagent negotiation (cf. Section 3.2) are a technology rise to implement autonomous industrial processes (cf. Section 3.3). Therefore, this chapter outlines multiagent-based applications in transport logistics and discusses their strengths and weaknesses in Industry 4.0 applications (cf. Section 3.4).

3.1 Intelligent Agents Constituting Multiagent Systems

A multiagent system consists of individual software agents. In general, there are multiple varying definitions of the term agent even in computer science. An intensive discussion about the term agent and its definition is, e.g., provided by Franklin and Graesser (1997) and Macal and North (2010). In this thesis, an intelligent and autonomous agent is defined as a software program which is an intelligent, virtual representative of an entity and/or provides a certain service. In the latter case, it is also called a service agent. Each agent has an objective which it tries to reach. A rationally acting agent selects "[...]an action that is expected to maximize its performance measure, given the evidence provided by the percept sequence and whatever built-in knowledge the agent has" (Russell and Norvig, 2010, p. 37).

There are several agent architectures which determine the actions that have to be performed in a given situation (or rather after a sequence of perceptions). Frequently referenced and well established agent structures are the simple reflex agent (Russell and Norvig, 2010, pp. 48-50), the model-based reflex agent (Russell and Norvig, 2010, pp. 50-52), the goal-based agent (Russell and Norvig, 2010, pp. 52-54), the utility-based agent (Russell and Norvig, 2010, pp. 54-56), the learning agent (Russell and Norvig, 2010, pp. 56-58), and mixtures of these architectures.

Within the agent's decision-making process its autonomy is the most important aspect. The term autonomy implies "[...]that agents are able to act without the intervention of humans or other systems: that agents have control both over their own internal state, and over their behavior" (Wooldridge, 2013, p. 5). Thus, an agent acts self-contained, is responsible for its decisions, and has complete control over its behavior and internal state. By its own decision-making and independent choice of performed actions, it achieves autonomy. An autonomous agent observes and perceives its environment. Thus, it has the capability to react to unexpected events (Brooks, 1986). By accumulated experiences and continuous learning it is able to adjust its behavior to new situations. Simultaneously, it operates proactively to reach its long-term goals (Rao and Georgeff, 1991, 1995).

In a multiagent system, the agents communicate and negotiate to determine an adequate mutual solution. Through cooperation between single agents on the micro-level, the emergent system behavior ideally leads to global optimization on the macro level. Wooldridge (2009, p. 26) describes the interplay between cooperation, coordination, and negotiation as social activities which are characterized by proactive, reactive, and adaptive behavior.

Similar to Müller (1997) who advises to apply multiagent systems in complex, dynamic and distributed problems, Jennings and Wooldridge (1998, pp. 5-10) suggest to implement multiagent systems especially in open and unpredictable environments in which it is not possible to find adequate solutions with predefined behavior. In addition, they argue that implementing multiagent systems is even more suitable in domains in which agents are a *natural metaphor*[1] of objects, in

---

[1]They refer to objects or actors of a physical system which can directly be mapped to agents in the virtual system.

which confidential information and data are stored decentralized with restricted user access, and in which participating parties have conflicting and/or selfish interests (Jennings and Wooldridge, 1998, p. 7).

In general, there are cooperative and competitive agent systems. In cooperative solutions the agents archive a global overall goal or at least a common sub-goal. For instance, several agents might represent employees of a logistics service provider. In such a cooperative scenario, the goals of individual agents are not relevant. The only goal is profit maximization for the enterprise, and all agents are committed to reach this goal. In more complex scenarios, the selfish, probably conflicting interests of single employees might also be considered within agent negotiations. Thus, the cooperative scenario is shifted to a competitive one.

A comprehensive survey of multiagent systems is provided by, e.g, Ferber (1999), Wooldridge (2009) and Weiss (2013). The wide range of multiagent systems' applications are discussed and concluded by, e.g.,Müller (1997), Jennings and Wooldridge (1998), and Kirn, Herzog, Lockemann, and Spaniol (2006).

3.2 Multiagent-Based Negotiations

In multiagent systems, semantic technologies and approaches, such as domain-specific ontologies, communication protocols, and speech acts (Austin, 1975; Searle, 1969), are applied to ensure the unambiguous communication between agents. In 1996, the *Foundation for Intelligent Physical Agents*[2] (FIPA) was established to specify standards for agent-based interactions and communications, e.g., agent communication languages, content languages, and architectures of multiagent systems. The goal of FIPA is to optimize the communication and interoperability in industrial and commercial processes.

In order to standardize the communication processes, protocols play a major role. They consist of ordered and defined sequences

---

[2]see: http://www.fipa.org (cited: 1.9.15).

of communicative acts, which are based on the speech-act theory of Austin (1975) and Searle (1969). Thus, communication protocols describe state-controlled communication processes which define "[...]possible actions that agents can take at different points of the interaction" (Sandholm, 1999, p. 201). In this context, a strategy is the mapping from state to action which determines the action an agent performs at a certain state of the negotiation. Consequently, selfishly acting agents follow a strategy which maximizes their individual profit. They are not influenced by other agents (Sandholm, 1999, p. 207).

The scientific area of *Mechanism Design* investigates the development of protocols which ensure predefined properties if (selfishly acting) agents follow a given strategy. A *mechanism* is a protocol for mutual decision-making within a system which consists of individual participants. Thus, the area is also related to game theory which was originally developed by Neumann and Morgenstern (1944). For instance, similar to game theory the expected *utility value* indicates the agent's *happiness* with a certain solution. There are several criteria for the evaluation of mechanisms which partly rely on concepts derived from game theory such as *dominant strategies*,[3] *Pareto efficiency*,[4] and *Nash-equilibria*.[5] Sandholm (1999, pp. 202-204) proposes the following criteria for the evaluation of mechanisms/protocols:

- Social welfare: The goal of a protocol which maximizes social welfare is to maximize the global overall utility of all agents. Nevertheless, the utility of single agents might be low in a solution which maximizes social welfare.

---

[3]A dominant strategy is the agent's i best strategy which maximizes his expected utility and which is independent of the strategies of other agents (Shoham and Leyton-Brown, 2009, p. 264).

[4]"A solution x is Pareto efficient - i.e. Pareto optimal - if there is no other solution x' such that at least one agent is better off in x' than in x and no agent is worse off in x' than in x"(Sandholm, 1999, p. 202).

[5]A solution is a Nash-equilibrium if all agents chooses their best strategy: "S_i^* is the agent's [i] best strategy - i.e. best response - given that the other agents choose strategies $(S_1^*, S_2^*, ..., S_{i-1}^*, S_{i+1}^*, ..., S_{|A|}^*)$" (Sandholm, 1999, p. 203).

- Pareto efficiency: In a Pareto efficient solution, no agent can increase its utility without decreasing the utility of all other agents.

- Individual rationality: The participation in a negotiation is individually rational if it has no negative effect on the agent.

- Stability: A mechanism is stable if it is not possible to manipulate the outcome of a negotiation for its own purposes.

- Distribution/effort of communication and computation: By problem decomposition, the computational complexity is reduced. Consequently, this increases the communication effort. The criteria evaluates the proportion between computation and communication efforts.

Likewise, Rosenschein and Zlotkin (1994) define similar evaluation criteria:

- Efficiency: This criteria is similar to the Pareto efficient criteria described above.

- Stability: The authors categorize a mechanism as stable if no participating agent has any incentive to dissent from an agreement (this criteria is similar to a Nash-equilibrium).

- Simplicity and Distribution: This criteria is similar to the distribution/effort of communication described above.

- Symmetry: A mechanism is symmetrical if no agent with the same properties and capabilities as other agents is discriminated.

As some of the criteria presented are qualitative ones, both authors leave it open how to measure these.

3.2.1 Auctions

Auctions and votes are familiar and established mechanisms for mutual decision-making. Auctions are performed to deal with products, services, and other resources. The participants try to purchase one or several goods. Their bids depend on how high they value the goods. An *efficient* auction ensures that the bidder with the highest bid (measured by its individual valuation of the goods) is the winner of the auction (Sandholm, 1999, p. 213). The valuation of goods could be either a *private value* (the individual value depends only on the rating of the participant), a *common value* (the goods have a factual value independent of the participant's rating) or it might be partially correlated by the value of other bidders (*correlated value*). Moreover, auctions differ in further aspects:

- The bidding rules define the sequence at which participants are allowed to submit offers. They specify the number of updating bids and determine the maximum or minimum amount of the new bids.

- The winner determination rules define the end of an auction, determine the participant(s) who purchase one or more of the auctioned goods, and define the price the winner(s) has/have to pay for it.

- Privacy aspects; in some auctions the bids of other bidders are visible for all other bidders. In other auctions bids of other bidders are kept under wraps.

Well established auctions are the English-auction, the Dutch-auction, the first-price sealed bid auction, and the second-price sealed bid auction also known as the *Vickrey*-auction (Vickrey, 1961). All of these auction types result in the same revenue for the auctioneer if the bidders have a private and independent valuation of the auctioned goods and act risk-neutral (Shoham and Leyton-Brown, 2009, p. 323). However, the situation is different for the bidders. While neither in the Dutch-auction nor in the first-price sealed bid auction, there is a dominant strategy, in the Vickrey auction "[...] where bidders have independent private values, truth telling is a dominant strategy" (Shoham and Leyton-Brown, 2009, p. 319). Similarly, truth telling is the dominant strategy in the English-auction, because it is strategically equivalent to the Vickrey auction (Sandholm, 1999, p.

213). However, the Vickrey auction is more relevant in application, because the communication effort is lower.

Common auction types for the allocation of multiple goods are combinatorial auctions, parallel auctions, and sequential auctions. In combinatorial auctions, all bidders submit their bids for all bundles of goods which they would like to purchase. The auctioneer evaluates all submitted bids and determines the winner by choosing the subset of bids which maximizes his profit. Thus, in combinatorial auctions all positive and negative synergies between goods are considered. However, the computational complexity is high: Firstly, the bidders have to compute the bids for all bundles (the number of computations grows exponentially to the number of goods). Secondly, the winner determination problem is even more complex, because the possible combinations of offers increase exponentially to the submitted bids. Although there are adequate heuristics, the computational complexity remains high in contrast to sequential and parallel auctions (Koenig, Tovey, Lagoudakis, Markakis, Kempe, Keskinocak, Kleywegt, Meyerson, and Jain, 2006).

In parallel auctions, each participant submits an independent valuation for each good in parallel. The auctioneer determines the highest bid for each item and accepts it. Therefore, the computational and communication effort is low and the auction is highly parallelized. However, synergies between goods are neglected.

In sequential auctions, the auctioneer provides one item after the other in several single item auctions until all goods are sold. Consequently, the bidders' valuations are based on the currently auctioned good as well as on the items they bought in preceding rounds. Thus, synergy effects between these goods are considered. Let O denote the set of goods and A the set of agents participating in an auction. The advantage of a sequential auction is that the computation and communication effort is limited, because each participant has to compute $|O|$ bids and the auctioneer has to compare $|O| \times |A|$ offers in total. However, not all relevant synergies between goods are considered within the agents' valuations. The outcome depends on the sequence in which goods are provided by the auctioneer.

Koenig et al. (2006) give a detailed theoretical investigation of the auction types described. Moreover, Krishna (2009) provides a comprehensive survey of auction theory.

3.2.2 The Contract-Net Protocol

The well-established and often applied *contract-net* protocol (Smith, 1977, 1980) is a generic protocol for multiagent-based negotiations. The general idea of a contract-net is to divide a problem into smaller sub-problems which are solved by single agents. Each agent has its individual properties and cost function, which defines the agent's effort to solve a set of delegated tasks. Initially, there might be a randomized allocation of tasks to agents. This non-optimal allocation is improved in continuous negotiations with other agents. Thus, a task is changed from agent i to agent j, if agent j solves a tasks at lower costs than agent i. The resulting allocation is approximated to an optimal solution, which maximizes the social welfare (Shoham and Leyton-Brown, 2009, p. 27).

The contract-net is classified as an *anytime* algorithm. Thus, the optimization process may be stopped before the optimal solution is reached and the best, valid solution found is returned (e.g., if a maximum computation time is reached). Especially in dynamic and stochastic environments, the available running times are not known in advance. Thus, an anytime behavior of the algorithm is a relevant property.

In 2002, the contract-net protocol was standardized by FIPA (Foundation for Intelligent Physical Agents, 2002). Figure 3.1 depicts a contract-net protocol, in which the initiator agent offers a task to m participating agents. These m agents compute their individual costs and send either a positive answer (a *propose message*) or negative answer (a *refuse message*) back to the initiator. The proposals of l agents are accepted while k agents receive a rejection of their proposal. Each agent that auctioned a task has to send an *inform message* back to the initiator if the task is successfully accomplished. Otherwise, it has to report a fault.

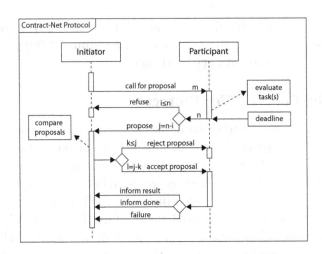

Figure 3.1: The contract-net protocol for agent-based negotiations (adapted from Foundation for Intelligent Physical Agents, 2002, p. 2).

While the contract-net defines the sequence of sent messages with specified performatives, it is open to the exact bidding rules and does not specify the agreements reached and their individual conditions. Thus, the protocol is general and can be applied universally in several application domains for different purposes. For example, the auctions presented in Section 3.2.1 may be implemented by the contract-net protocol. Since only two agents are involved in a communication, the negotiation is also called *swap-contract*. Since a bundle of tasks is auctioned at the same time this type of the contract-net is named *cluster-contract* (Shoham and Leyton-Brown, 2009, p. 28).

3.3 Multiagent-Based Autonomous Logistics

In autonomous logistic processes, the conventional planning and control is reversed and shifted from centralized, hierarchical systems to decentralized, heterarchical systems (Scholz-Reiter, Windt, Kolditz, Böse, Hildebrandt, Philipp, and Höhns, 2004), in which intelligent

logistic objects play a major role: "Autonomous control enables logistics objects to process information, to make and execute decisions, and
to cooperate with each other based on objectives imposed by their
owners" (Schuldt, Hribernik, Gehrke, Thoben, and Herzog, 2010, p.
2). Thus, the decision-making is delegated to single entities which
interact autonomously with other entities such as packages, vehicles,
robots, manufacturing plants, or essential parts of complex production
facilities.

However, an object itself is a *thing* which is acted upon, but not
able to act autonomously. For the implementation of autonomous
control, there is the essential requirement to transform an object
into an autonomously acting, intelligent entity. Therefore, logistic
entities are equipped with core technologies that transform the objects
into Cyber Physical Systems (CPS) with communication technologies
(e.g., wireless local area network), localization systems (e.g., satellite
positioning systems), identification technologies (e.g., radio-frequency
identification), data processing units (e.g., embedded systems), and
additional sensors and actors (Schuldt, 2011, pp. 44-63).

Multiagent systems (see Section 3.1) can be used to implement
autonomous control and to link the physical world with virtual representatives. In multiagent-based autonomous logistics, intelligent
software agents represent logistic entities, e.g., containers or vehicles.
Thus, they are able to autonomously plan and schedule their way
throughout the logistics network (Schuldt, 2011, p. 95). The agents
act on behalf of the represented objects and try to reach the objectives
assigned to them by their owners. Consequently, relevant information
is directly linked to products. For instance, an agent representing a
shipment is aware of its individual weight, volume, and its designated
place and time of arrival. The material flow is directly connected to the
information flow which allows agents to receive and process all relevant
data immediately. Consequently, relevant data need not be filtered
out of big data stores. Considering real-time information about the
status of the physical world, the quality of the agents' decision-making
processes is improved. In addition, process disturbances are firstly
handled locally without effects to the overall system. In addition to

this local optimization, agents interact to further optimize the overall system performance. For instance, in case of a traffic congestion, an intelligent agent representing a vehicle ideally determines alternative routes autonomously to reach the next stop on time. If this is not possible, it might alter the stop sequence or pass orders to other agents to ensure reliable transport times. In these interactions, the agents share their knowledge by communication and negotiation mechanisms with other agents, in order to optimize process efficiency and resource utilization (see Section 3.2). Moreover, by delegating planning and control processes to decentralized entities, the overall problem is split into smaller problem instances. The computational complexity of the resulting problems is reduced. In many cases, these reduced problems can be solved optimally.

In conclusion, the decentralization of planning and control as well as the problem decomposition decrease the computational complexity and increase the system's flexibility and adaptability in highly dynamic environments. A survey of autonomous control in logistics is provided by Hülsmann, Scholz-Reiter, and Windt (2011).

3.4 Multiagent-Based Systems in Transport Logistics

In general, centralized solvers dominate logistics software solutions which are rooted on mathematical methods and Operations Research like these of Dumas, Desrosiers, Gelinas, and M. (1995), Dorigo and Gambardella (1997), Bräysy et al. (2004), Bräysy and Gendreau (2005a), Bräysy and Gendreau (2005b), Homberger and Gehring (2005), to name but a few (cf. Chapter 2). Nevertheless, there are also several multiagent-based approaches for resource allocation and scheduling problems, which are described in this section. This investigation concentrates on transport scheduling and planning problems, e.g., on the allocation of transport tasks to vehicles, fleet management, and routing problems. While this class of problems refers to problems of transporting goods from origin to destination, multiagent-based approaches for other logistics domains such as supply chain management, traffic management and control, train and air scheduling, in-house logistics, hospital logistics, or container terminal management, are neglected. Solutions for in-house logistics are, e.g., provided by Ten Hompel (2006) or Lewandowski et al. (2013). A multiagent-based approach for patient scheduling in hospitals is given by Braubach, Pokahr, and Lamersdorf (2013b). A survey of multiagent-based approaches in logistics in general is provided by, e.g., Davidsson, Henesey, Ramstedt, Törnquist, and Wernstedt (2005), Pokahr, Braubach, Sudeikat, Renz, and Lamersdorf (2008), as well as Skobelev (2011).

As described in Section 3.1, agents represent logistic entities such as shipping companies, vehicles, containers, and shipments. In some cases, there are also agents which only provide services as answering planning and routing requests. Determining an adequate level for the representation of logistic entities by agents depends on several factors and the domain the system is designed for (Windt, 2008, p. 350). Consequently, if small-size objects or parts of objects are represented by their own agent, technological limitations have to be considered (Schuldt, 2011, pp. 42-44). For instance, "the more decision-making is distributed from one or few central to many local entities, the more communication is required for coordination" (Schuldt, 2011, p. 43). However, a more decentralized structure increases the number of concurrently running processes. This is especially relevant if the multiagent system is running on a cloud computing platform. In addition, fine-grained modeling ideally ensures that confidential data is not necessarily sent to and processed by other agents, while higher levels of modeling allow agents to have more relevant information available to improve the quality of their decisions. Generally, in approaches with fine-grained model agents directly represent single logistic objects, while in approaches with a coarse-grained model agents often represent organizational entities. Independent of the level of modeling in most approaches presented, the agents typically communicate, negotiate, and cooperate (or even compete) by applying the contract-net protocol (cf. Section 3.2.2) to reach their individual goals. Nevertheless, some authors additionally apply other bargaining protocols in certain applications (Fischer, Müller, and Pischel, 1996, pp. 12-14).

Firstly, Section 3.4.1 outlines and discusses approaches which implement coarse-grained representation levels. Next, Section 3.4.2 follows with the presentation and discussion of methods with fine-grained representation levels. Finally, Section 3.4.3 discusses all different approaches and draws a conclusion.

3.4.1 Approaches with Coarse-Grained Representation Levels

Table 3.1 gives an overview of approaches with a rather coarse-grained representation level which are discussed in this section.

Table 3.1: An overview of approaches with a coarse-grained representation level.

Published	Authors	cf. Page
1996	Fischer, Müller, and Pischel	48
2000	Bürckert, Fischer, and Vierke	49
2001	Thangiah, Shmygelska, and Mennell	50
2003	Perugini, Lambert, Sterling, and Pearce	51
2005	Dorer and Calisti	52
2012	van Lon, Holvoet, Vanden Berghe, Wenseleers, and Branke	53

Fischer, Müller, and Pischel (1996)

Fischer et al. (1996) have been among the first to present a flexible agent-based approach for complex resource allocation problems in dynamic environments. The goal of this approach is to efficiently allocate orders within a shipping company to its vehicles and simultaneously to change orders between several companies if it is more profitable to handle a task through another company. In this system, agents represent vehicles and shipping companies. To allocate tasks, the shipping company agent acts as a centralized auctioneer offering each task sequentially. Therefore, the agent initiates a slightly modified contract-net protocol (called *Extended Contract-Net Protocol* (ECNP)) to determine an appropriate vehicle for handling a task (Fischer et al., 1996, pp. 10-11), whereas in inter-company negotiations a bargaining protocol is applied (Fischer et al., 1996, pp. 12-14). A simulated trading algorithm for changing orders between vehicles analogous to a stock-exchange is developed to react to changing environmental influences and new order situations (Fischer et al., 1996, pp. 19-22). Consequently, the management and maintenance of the order-exchange require another centralized component. As numerous other multiagent systems for task and resource allocation problems, this system is sensitive to the sequence of incoming/dispatched orders and assumes that agents are totally cooperative in inner-company negotiations. The approach is evaluated with the instances of Solomon's benchmark set (Solomon, 1987) where all orders are known in advance. "It shows that the ECNP solution is between 3% and 74% worse than the optimal solution[...]"(Fischer et al., 1996, p. 24). Several of the following multiagent systems are based on the general approach and/or on concepts of Fischer et al. (1996).

Bürckert, Fischer, and Vierke (2000)

For instance, Bürckert et al. (2000, pp. 697-725) presented a multiagent system which is similar to the system of Fischer et al. (1996). They developed a so-called holonic multiagent system for planning and scheduling processes of a shipping company for real-world operations. In the context of the organization of agent societies, a holonic agent "[. . .] is a team of agents that committed themselves to cooperate and work towards a common goal in a cooperated way and look like an individual, coherent agent to the outside environment." (Bürckert et al., 2000, p. 707). In this system, trucks, trailers, drivers, as well as route planning and handling services are represented by agents that form collectives (namely holons). The head of a holon is a managing agent which is responsible for the administration of the holon. In addition, the head represents the holon in the agent's society. This head is the planning agent. The planning agent forms the holon and negotiates

with company agents that auction orders. In both negotiations, the
agents apply the ECNP developed by Fischer et al. (1996, p. 10-11).
Similarly, the system adapts the simulated trading algorithm and
other components of Fischer et al. (1996, pp. 19-22). Although the approach presented was transformed to a commercial application called
TeleTruck, neither comparison to other multiagent-based approaches
nor performance measurements in real-world application are provided.

Thangiah, Shmygelska, and Mennell (2001)

Similar to the classic approach of Clarke and Wright (1964) (cf. Section 2.2 on Page 24), Thangiah et al. (2001) provide a multiagent-based
approach to solve the Vehicle Routing Problem (VRP). A central
auction agent offers orders to vehicle agents sequentially (the ordering
is not described in the paper). The vehicle agents apply an insertion
heuristic to compute the additional cost for transporting the order.
As long as it is possible to transport the order, it is allocated to
the vehicle agent which has the lowest transport costs. Otherwise, a
new vehicle is created which starts a new tour with this order. After
the initial allocation process is finished, the continuous improvement
strategy is applied which allows vehicles to change single orders. The
approach is not implemented on an agent platform that satisfies the
FIPA standard but creates its own non-standardized communication
protocols.

Although the system is compared to other state-of-the-art Operations Research solvers, it is not mentioned which benchmark is
applied. The results show that the multiagent-based approach is not
competitive with state-of-the-art solutions. The authors assume this
is caused by the application of "a very simple heuristic [. . . instead of]
powerful algorithms based on Genetic, Tabu and Simulated Annealing
Metaheuristics" (Thangiah et al., 2001, p. 520). The main advantage
of the system is that it can easily be extended to solve multi-depot
problems as well (Thangiah et al., 2001, p. 520).

Perugini, Lambert, Sterling, and Pearce (2003)

Perugini et al. (2003) developed a multiagent system to support the logistics in military planning. They model two types of agents. On the one hand, there are managing agents that represent selfish organizational entities, which have to manage specific logistic tasks. On the other hand, there are service provider agents that offer diverse transport services, e.g., transport by plane or cargo ships. The agents act selfishly and have to conceal confidential data from other agents. Thus, Perugini et al. (2003) argue that their multiagent system is promising to solve this military problem, because their agents keep extensive information private. They only reveal their available capabilities when they are required to perform the task. As a managing agent has to schedule the transport of a large amount of goods or even a complete military unit, this agent has to allocate the goods to multiple service providers. In addition, the problem changes to a multimodal problem. For instance, it is necessary to transport the military equipment by cargo ships first, while trucks carry the goods to their final destination. This differs from the problems investigated in this section, in which the agents have to find a single transport service provider for the overall transport for each item. Thus, the additional combinations significantly increase the computational complexity of the problem.

Therefore, Perugini et al. (2003) adapt the ECNP of Fischer et al. (1996, pp. 10-11) to additionally support partial routes in the negotiations. Moreover, they allow backtracking: The original protocol only permits rejecting all temporally granted bids, while the newly developed *Provisional Agreement Protocol* enables the managing agent to reject a single bid only (Perugini et al., 2003, pp. 20-21).

The protocol supports combinational auctions to solve a complex logistics problem. However, the mechanism does not guarantee optimal solutions by considering all combinations of auctioned items. The authors implemented and evaluated the protocol only in a small scenario with a single managing agent and three transport service providers. Neither a theoretical analysis of the mechanism nor another evaluation is provided which allows to judging the efficiency and solution quality. Although privacy aspects are the motivation for applying a MAS, the managing agents still have to reveal information about the transported goods to transport service providers and these have to announce their available capacities. Nevertheless, the approach profits from applying a multiagent system, because not all the information is centrally available. For example, the managing agents are unaware of each other's requests and the transport service providers have no knowledge of each other's available capacities.

Dorer and Calisti (2005)

In order to reduce the communication effort and to increase the scalability, Dorer and Calisti (2005) follow the divide and conquer strategy to split up the catchment area in multiple smaller sub-areas which are controlled by a single managing agent and an order agent. The managing agent sequentially receives the transport requests from order agents and tries to insert each request to one of the tours of his controlled area. Therefore, the agent applies a cheapest insertion heuristic, which determines the costs. During the agent's decision-making process several domain specific hard and soft constraints are considered (Dorer and Calisti, 2005, p. 46-47). It is also possible that an allocated order cannot be transported after all because of its lower priorities. Consequently, this order is processed for a second time. If the insertion fails again, the order is marked as unprocessed. In case the order's pickup and delivery location are scattered over several sub-areas, the respective managing agents communicate with each other. As a result, the sub-areas should not overlap and should optimally include closed districts such as a whole city. Otherwise, the divide and conquer strategy increases the communication effort and turns the advantage into a disadvantage.

The approach is evaluated by a simulation of a real-world-scenario with about 3,500 orders in cooperation with Whitestein Technologies' Living Agents Runtime System.[6] The resulting plan "[...[was checked

---

[6]see http://whitestein.com for more information (cited: 1.9.15).

by dispatchers for feasibility and drivability. [...] A total of 11.7% cost savings was achieved, where 4.2% of the cost savings stem from a corresponding reduction in driven kilometers." (Dorer and Calisti, 2005, p. 51).

van Lon, Holvoet, Vanden Berghe, Wenseleers, and Branke (2012)

van Lon et al. (2012) present a multiagent system to solve the Dial-a-Ride Problem (DARP) (cf. Chapter 2). The modeling is rather coarse-granular, because agents only represent vehicles. The approach is fully decentralized and agents make their decisions by themselves. The agents have full knowledge about all orders but have a restricted proactive behavior because they plan only a single order in advance: Each vehicle agent receive requests from order agents and computes a priority value for each request by a genetic algorithm. The algorithm considers the individual properties of the agent and these of the orders to compute the priority value. Next, each agent processes the request which has the highest priority. As a result, conflicts may arise if more than one agent accepts the same request. In this case, the first agent to arrive at the pickup location prevents the others from processing this request. Consequently, no communication is required between agents, but some agents have unnecessarily started driving to the pickup location. This is a significant lack of optimization. The authors evaluate the system by simulation of an artificial scenario in the real-world infrastructure of the city of Leuven (Belgium) and show the effectiveness of the approach.

3.4.2 Approaches with Fine-Grained Representation Levels

Table 3.2 gives an overview of approaches with a rather fine-grained representation level which are discussed in this section.

Table 3.2: An overview of approaches with a fine-grained representation level.

Published	Authors	cf. Page
1999	Kohout and Erol	54
2006	Leong and Liu	55
2007	Mes, van der Heijden, and van Harten	56
2008	Barbucha and Jędrzejowicz	57
2009	Zhenggang, Linning, and Li	58
	Magenta Technology:	59
2005	Himoff, Skobelev, and Wooldridge	
2006	Himoff, Rzevski, and Skobelev	
2009	Glaschenko, Ivaschenko, Rzevski, and Skobelev	
2010	Máhr, Srour, de Weerdt, and Zuidwijk	63
2010	Vokřínek, Komenda, and Pěchouček	66
2012b	Kalina and Vokřínek	

Kohout and Erol (1999)

Kohout and Erol (1999) modeled vehicles and orders as agents which negotiate by applying the contract-net protocol (cf. Section 3.2.2). Indeed, the presented multiagent system is a parallel multiagent-based implementation of Solomon's algorithm (Solomon, 1987). Tasks are initially allocated sequentially in the sequence in which they are presented to the system. Each available vehicle is asked for the cost for transporting a new incoming task. If the order agent receives at least one positive answer, the task is inserted in the lowest cost route. Otherwise a new vehicle agent is created which initializes a new route with a single task. As with to Solomon (1987), this algorithm is based on the *Savings Heuristics* of Clarke and Wright (1964). After allocating the available orders to vehicles, an improvement phase is started. The vehicle agents select randomly orders from their tours and offer these orders to other agents, who might service the order with less cost. This process is continued until a deadline is reached. Thus, the process emulates a 2-opt-exchange improvement technique (Lin, 1965a). In order to determine the costs for transporting a task, an algorithm similar to Solomon's I1 insertions heuristic (Solomon, 1987) is applied in the agent's decision-making process. The evaluation showed that the concept of iterative order negotiations yields to reasonable results. However, the system is more time consuming than the centralized version and was only tested on static test instances which have been adapted from the well-known Solomon benchmark set. As the adopted benchmark set varies from the original one, it is impossible to compare the performance with other approaches. Although the system is tested on static problems, it is designed for solving dynamic problems as well.

Leong and Liu (2006)

Leong and Liu (2006) present a combination of a central local search algorithm computing an initial solution first, and a multiagent system improving this solution afterwards. The initial solution is computed by Solomon's insertion heuristic (Solomon, 1987). Next, three kinds of agents are created: customer agents, route agents, and a planner agent. Customer and route agents optimize their objectives locally. Therefore, they compute so-called *moves* to optimize the tour as well as the route, which can be executed by the agents. All computed moves are sent to the planner agent. The planner agent maintains a *move pool*(Leong and Liu, 2006, p. 107). In addition, the agent has all information about the customers and current routes. It optimizes the tours globally by applying a 2-opt exchange heuristic (Leong and Liu, 2006, p. 110) and eliminates so-called *bad routes* to reduce the total number of routes. Moreover, the planner agent selects the *best moves* and sends its decision back to the other agents, which have to execute the moves. As moves in the move pool are eventually affected by selected decisions, this is not detected. Instead, the correctness of best moves is checked before they are selected. However, this may affect the optimization process of the planning agent and result in further invalid proposals. The developed multiagent system contains only cooperative agents. All agents (except for the planning agent) have to execute the decisions made by the central planning agent. At the same time the central planning agent acts as a centralized solver which must have all the information about other customers and routes. Thus, the system could be described as a decentralized implementation of a centralized approach. The agents act as information providers to the decision maker rather than making their own decisions. This perspective is strengthened because the system "[...] is not implemented using a *distributed* multi-agent system" (Leong and Liu, 2006, p. 110) and has no underlying agent architecture that satisfies the specification of FIPA. Instead, it is the centralized implementation of a multiagent concept. The system is evaluated with Solomon's benchmark set (Solomon, 1987).

Mes, van der Heijden, and van Harten (2007)

In contrast to other multiagent-based approaches presented in this section, Mes et al. (2007) consider a dispatching problem of (only) full-truckload freight with time windows. Thus, the delivery of each container follows immediately after the pickup. This reduces the planning complexity significantly. The orders appear dynamically during operations. The authors chose a fine-granular model, in which each order and vehicle is represented by its own agent. The agents apply the Vickrey auction (cf. Section 3.2.1) for negotiations. Once an order agent is created, it starts a new auction to find a proper transport vehicle. In order to determine their additional costs, the vehicle agents have to update their tours. For tours containing less than ten orders, the resulting TSP is solved by a branch-and-bound algorithm, which computes lower bounds by an insertion algorithm (Mes et al., 2007, p. 66). As new orders might alter the solution quality of already computed tours, vehicles may change orders between each other. Unfortunately, this changing process is limited to the following transfer: "Whenever a vehicle, after unloading at a certain terminal i, has to travel empty to terminal j, its agent searches for another vehicle agent that has a job from i to j that has been released but that is not started yet. Then the job that yields the highest savings (if positive) will be transferred to the vehicle to avoid empty traveling" (Mes et al., 2007, p. 66). However, this is a rather rare case in real-world processes. For the evaluation, the authors compare their multiagent system to two centralized approaches of Mes et al. (2007, p. 67-68). Therefore, they implement an artificial scenario into a small manually modeled infrastructure (of Amsterdam) as well as into a small randomly generated road network (Mes et al., 2007, p. 68-69). The results show (Mes et al., 2007, p. 71-74) that the developed multiagent system "yields a high performance in terms of vehicle utilization and service level" (Mes et al., 2007, p.74). However, the system is only compared to two other centralized systems, whose efficiency and quality of computed solutions are not indicated, in these self-modeled artificial scenarios.

Barbucha and Jędrzejowicz (2008)

The goal of Barbucha and Jędrzejowicz (2008) is to combine an established centralized approach with a multiagent system to solve dynamic VRPs. Firstly, all orders are allocated to vehicles by the algorithm originally developed by Gillett and Miller (1974). Next, dynamically incoming orders are sequentially auctioned by a central so-called *request manager agent* (Barbucha and Jędrzejowicz, 2008, p. 519). The agents use the contract-net protocol to negotiate (cf. Section 3.2.2). The vehicle agents compute the additional costs for handling the new order by its cheapest insertion between two stops of the current tour. However, there is no continuous improvement strategy. Thus, the optimization potential is decreased significantly by restricting the search space. The algorithm converges in a local maximum, and possibly valid solutions are not found. If there is no vehicle which can transport the order without violating capacity constraints, the order is refused and remarked as unserved. In adapted benchmarks, the authors evaluate the performance in environments with varying dynamics. This allows investigating the robustness of the multiagent system in dynamic environments, but it precludes the comparison with other approaches.

Zhenggang, Linning, and Li (2009)

Zhenggang et al. (2009) investigate the impact of reducing the number
of negotiations started by a central scheduling agent, which auctions
incoming tasks sequentially to vehicle agents. While in other approaches the orders are offered to each vehicle, in their system only
a subset of vehicles is considered in negotiations. As a result, the
communication effort is reduced, but also potentially good allocations
are not found by this strategy. To determine vehicle agents who are
not allowed to participate in a negotiation, one criteria of Zhenggang
et al. (2009, p. 411) is the distance of the pickup and delivery location
of a new task to existing stops on the route. Thus, only nearby
vehicles receive a *call for proposal* message. However, this might mean
that an order, which is directly on the route of a vehicle, has to be
processed by another vehicle if the stops are not near to the currently
auctioned order. Similar to the multiagent system of Barbucha and
Jędrzejowicz (2008), no continuous improvement strategy is applied.
Thus, the solution quality is highly sensitive to the sequence with
which the orders are allocated by the scheduling agent. However,
a resulting advantage is that their approach allows neglecting all
vehicles in negotiations which have reached their capacity limits. The
multiagent system is only evaluated by the sub-set C1 provided by
Solomon (1987). The results show that the number of negotiations can
be reduced significantly without any impact on the solution quality.
However, the C1 sub-set contains exclusively problems with clustered
structures. In addition, each vehicle has to transport orders of one or
at least two nearby clusters in the solutions (Zhenggang et al., 2009,
p. 411). It is evident that in this particular problem the vehicles
which are located in other clusters can be neglected in negotiations.
Nevertheless, it is remarkable that the approach determines adequate
solutions without any improvement strategy. This is probably due
to the orders being auctioned in the sequence of their numbering
and that each tour (*solution cluster*) contains orders with successive
numbers in the considered problem instances (cf. the figure provided
by Zhenggang et al., 2009, p. 411). For instance, in the example

examined by Zhenggang et al. (2009, p. 411) orders are successively allocated to the same vehicle as long as they are located nearby. If it is not possible to transport the order by the vehicle (because of capacity or time constraints), the next vehicle agent purchases the order by auction. Thus, this vehicle agent is responsible for the next cluster and all will auction the following orders as well. This process is continued until all orders are allocated. The example pinpoints that the approach is highly sensitive to the sequence in which orders are allocated to vehicles. Thus, changing the initial numbering of orders or applying the randomized sub-set R1 of Solomon's benchmark, would probably lead to significantly worse results.

Magenta Technology: Himoff, Skobelev, and Wooldridge (2005); Himoff, Rzevski, and Skobelev (2006); Glaschenko, Ivaschenko, Rzevski, and Skobelev (2009)

The *i-Scheduler*, which is based on *Magenta Agent Technology*[7], was originally developed for the scheduling of about 46 oil tankers of a shipping company called Tankers UK Ltd (Himoff et al., 2005, p. 60). *Magenta Agent Technology* is a framework which provides an ontology-based knowledge base and tools for the implementation of multiagent systems. The framework applies the so-called *Virtual Market Energy* as underlying agent platform (which is not FIPA-compliant). It allows the modeling of agents, roles, and message types which are required for negotiation (Himoff et al., 2005, p. 63). The special feature of Magenta's agent technology is that it automatically transforms manually modeled entities and multiagent systems to executable Java code. As a result, domain experts can ideally model multiagent-based applications by graphical user interfaces themselves. The *i-Scheduler* is a multiagent system in which agents negotiate and swap orders between each other to find adequate solutions for a real-world transport problem. Unfortunately, Himoff et al. (2005) describe neither the implemented negotiation mechanisms nor the decision-making processes in detail. On the one hand, the multiagent

---

[7]For more information see http://magenta-technology.com/ (cited: 1.9.15).

system is applied in real-world applications. On the other hand, no further evaluations, performance measurements, or comparisons to other scheduling approaches are provided.

Himoff et al. (2006) extended the software for the scheduling of road transportation. Their goal is to apply the system in dynamic environments and especially to handle the uncertainty of planned transport schedules. Thus, the system is designed to react to events as they appear (Himoff et al., 2006, p. 1519). The most important agents are following three types: The order agent "splits orders into transportation instructions (TI) and monitors its Key Performance Indicators" (Himoff et al., 2006, p. 1518). The *transport instruction agent* "searches for the best journeys" (Himoff et al., 2006, p. 1518). The *journey agent* "looks for the best TIs, devises good consolidation, [and] looks for the best routes with minimal mileage" (Himoff et al., 2006, p. 1518).

Similarly to the approach of Bürckert et al. (2000, pp. 697-725) (cf. Page 49), agents form coalitions, e.g., in order to bundle several TIs to form a tour. A single agent represents a created coalition. Other agents may join and leave a coalition dynamically. The authors do not reveal the applied tour and routing algorithms, which are an essential part of the agent's decision-making process. Neither they do disclose the concrete communication acts of the negotiation protocols and how coalitions are formed in detail (especially the conditions for entering or leaving a coalition - they only refer to "cost"). For instance, it is mentioned, that negotiation branches are prioritized and *uninteresting options* are detected and cut negotiation branches early. However, it is not revealed how the branches are prioritized and how *uninteresting options* are defined or detected. Instead, the system is very generic in order to support individual customer objectives and criteria, e.g., by a weight coefficient.

A key feature of the system is that it reacts to dynamic events. For instance, if a new order appears, the system first tries to insert this new order in the current plan. Therefore, it has to interrupt all concurrently running processes and negotiations. If several orders have to be included at the same time, the systems inserts each order sequentially - one after the other. Concurrent insertions of orders are not supported even if plans are not related to each other. For example, the system does not allow changing two tours of different catchment areas at the same time.

The multiagent system is evaluated by real-world data of an industrial partner. The results reveal that only 5% of all computed routes have to be rescheduled manually. In the system currently applied by the industrial partner, about 30% of computed tours have to be rescheduled. No objective performance measurements are provided which would allow comparing or evaluating the quality of computed solutions.

Glaschenko et al. (2009) further extended the system for real-time dynamic scheduling of taxi companies. Therefore, they extended the multiagent system to handle individual requirements of the taxi domain. For instance, automated planning makes it possible to consider the favored payment methods, child seat if required, and smoking preferences.

However, compared to a standard VRP, the complexity of this dynamic scheduling problem is lower, because each vehicle has a limited capacity of exactly one (similarly to Mes et al. (2007)). Thus, it is not possible to bundle orders. Consequently, this also decreases the complexity of the agents' decision-making processes significantly. Thus, each order can be processed in a sequential way and there are neither positive nor negative synergies between orders, which facilitates the planning processes. The authors assume there are enough vehicles to service all the requests. It is not described how the system handles unserved service requests. For instance, it is not possible to shift requests with lower priority to other time slots later on. The applied (routing) algorithms are not described, although they are essential for the agents' decision-making.

Nevertheless, the orders have differing priorities. Thus, the vehicles prefer orders with high priority of premium customers. This also leads to the reallocation of previously allocated orders (Glaschenko et al., 2009, p. 32). "The length of the re-scheduling chain is limited only by the time required to reach a client in the busy city of London, which normally is sufficient for several changes of the schedule" (Glaschenko et al., 2009, p. 32). Further details about the reallocation are not provided, e.g., how many reallocations are performed on average and if the reallocation process avoids infinite loops. In addition, it could make sense to continue the negotiations. For instance, if a new vehicle enters the system in the neighborhood of a customer, changing orders could increase the solution quality. In general, agents do not act selfishly but cooperate with each other. Thus, it is irrelevant that an agent might lose an order without receiving another one instead. Cheating of drivers is avoided by the application of further artificial intelligence and the surveillance of the drivers (Glaschenko et al., 2009, p. 33). For example, the system automatically detects cheating drivers who obviously insert incorrect data about their position or status.

In order to reduce the interaction effort, the authors split up the overall catchment area into several smaller sub-areas. This avoids considering vehicles located in other sub-areas far away from the customer. The principle of the sub-areas relies on the assumption that each order may be changed as long as there is enough time for the taxi to reach the customer. Consequently, another pattern-recognition approach forecasts upcoming order situations and determines a reasonable distribution of free taxis to sub-areas (no more detailed information is provided). The approach was evaluated by relatively small problem instances (with less than 100 orders) in real-world applications (Glaschenko et al., 2009, p. 32). The results show that 98.5% of all tours could be computed automatically. In addition, "the number of lost orders was reduced to 3.5 [%] (by up to 2 %); the number of vehicles idle runs was reduced by 22.5%. Each vehicle was able to complete two additional orders per week spending the same time and consuming the same amount of fuel, which increased the yield of each vehicles by 5-7 %" (Glaschenko et al., 2009, p. 34). Further evaluations or comparisons to other approaches are not provided.

Máhr, Srour, de Weerdt, and Zuidwijk (2010)

In order to strictly avoid central agents, Máhr et al. (2010) modeled each order by its own agent. In their multiagent system, order agents immediately start negotiating with vehicle agents when they appear in the system. Consequently, vehicle agents might participate in several auctions concurrently. This requires additional coordination to handle possible conflicts in the bidding process. For instance, an offer might be out of date because another contract has been accepted concurrently. The resulting problem is known as the Eager Bidding Problem (Schillo, Kray, and Fischer, 2002). Máhr et al. (2010, pp. 7-8) solve this problem by updating the costs for accepted proposals which were sent before the plan was changed but accepted afterwards. If the costs are equal or lower, the negotiation is continued. Otherwise, the vehicle agent interrupts the negotiation and the order agent starts a new one.

In contrast to other presented approaches, Máhr et al. (2010) apply the Vickrey auction (cf. Section 3.2.1) instead of the contract-net protocol to communicate. As a result, the winning agent is aware of the valuation of the second best offer and can use this information in its decision-making process as described below. To compute the costs for servicing an order, the authors apply a combination of an insertion algorithm (which is based on Solomons I1 heuristic (Solomon, 1987)) and a so-called substitution algorithm. The substitution algorithm determines for each existing order the costs of substituting it by the new order. In this process, a substitution penalty is considered. Its amount depends on the high of the payment the vehicle has to pay for the substituted order in the Vickrey auction. Therefore, the penalty is high if the payment was high which implies that the costs for transporting the goods by another vehicle would also be high. If the already integrated order with the lowest cost is cheaper than the insertion costs, this order is substituted by the new order. However, the substitution processes might result in a seemingly infinite loop of negotiations. To limit the number of negotiations, a substitution is only allowed if the new plan is better than the old one by at least ϵ. On the one hand, the substitution guides the search in new directions, because it increases the possibility to find new solutions by the removal and reallocation of an allocated order. On the other hand, this strategy might be insufficient, because the substitution is limited to a single shipment. For instance, the system does not find any solution if a new order can only be transported by a certain vehicle which has to remove two orders from its current plan (although these two orders might be transported by other vehicles).

For continuous improvement, each allocated order tries to reallocate itself at random intervals to find a vehicle that transports the order with less costs. However, the reallocation can result in further substitutions. As a consequence, the anytime behavior to improve a valid solution monotonically is not guaranteed. For example, a solution in which all orders are allocated might be changed to a suboptimal allocation, in which the costs are indeed less, but certain orders are not transported.

Beside this 1-opt improvement strategy, a k-opt improvement strategy is implemented. Therefore, a vehicle agent offers a segment of its plan to another randomly chosen vehicle agent (the considered plan segments contain three orders at most). The responder agent "performs a full search on the k-exchange neighborhood of the two plan segments, and send [sic!] back the best exchange combination to the initiator, if one is found that lead [sic!] to better plans." (Máhr et al., 2010, p. 11). As the total costs of both plans is considered, one of the agents might have higher cost than before. Thus, the improvement strategy requires cooperative vehicle agents. In addition, agents have to reveal a part of their plans. The k-opt improvement strategy is limited to the exchange of plan segments between two agents only. To prevent conflicts that appear if orders are loaded during their negotiation and reallocation process, the vehicle agents apply a *con-*

servative synchronization strategy[8], which ensures that goods whose order agents are engaged in negotiations are not transported but on hold. However, in this case the planning processes interfere and delay the operational processes. In the worst (but rather unlikely case) this results in a finite loop of disturbances. The probability of such disruptions is increased with the number of improvement procedures performed during operation.

In general, there are orders that temporarily cannot be transported by any truck. These orders have to start new negotiations once the situation changes. For this purpose each vehicle sends a notification to all orders if their plan has changed. This strategy has two disadvantages. Firstly, environmental changes, such as a delay or an accelerated transport, are not considered. Secondly, the number of negotiations is increased exponentially to the number of orders.

The goal of Máhr et al. (2010) is to compare a state-of-the-art multiagent-based approach to an established centralized algorithm in uncertain environments with varying degrees of dynamics (Máhr et al., 2010, p. 18). Therefore, they conducted a case study in cooperation with an industrial partner in the Benelux. However, they did not evaluate the multiagent system on any benchmark set or value the general performance of both implemented systems compared to other existing approaches. The results show, that their multiagent-based approach is highly competitive with their implementation of a state-of-the-art Operations Research algorithm if less than 50% of the orders are known in advance (Máhr et al., 2010, pp. 25-28).

---

[8] *Conservative synchronization* strategies prevent conflicts in advance while *optimistic synchronization* strategies detect conflicts as they arise and solve these by applying a roll-back (Jefferson, 1990, p. 75). The advantage of optimistic synchronization is that fast agents (processes) do not have to wait for slower agents (processes). The disadvantage is that the history has to be saved to ensure that roll-backs can be performed.

Vokřínek, Komenda, and Pěchouček (2010); Kalina and Vokřínek (2012b,a)

Vokřínek et al. (2010), Kalina and Vokřínek (2012a), as well as Kalina and Vokřínek (2012b) provide a multiagent-based solver for VRPs and PDPs with several constraints such as time windows and capacity limits. The focus of their investigations is on the comparison of the presented multiagent system with other established centralized approaches on static problem instances and on a comprehensive algorithmic analysis. Therefore, the system is evaluated with the well-known Solomon benchmark instances (Solomon, 1987), the benchmark instances of Homberger and Gehring (2005), and those of Li and Lim (2001).[9] Their multiagent system contains three classes of agents. The *task agent* is responsible for collecting all incoming orders and passing either a single task or a bundle of tasks to the central *allocation agent*. There are several strategies that determine the sequence in which tasks are sent, because the approach is highly sensitive to the ordering in which tasks are processed by the allocation agent. Next, the allocation agent starts a contract-net negotiation (cf. Section 3.2.2). Vehicle agents are responsible for the computation of routes. Therefore, they apply a cheapest insertion algorithm (Solomon, 1987). The allocation agent has two strategies to allocate a task to a vehicle agent:

1. If no vehicle is found which can handle a task, a new vehicle is created, all orders are returned to the task agent, and the procedure is started again.

2. The best vehicle found is forced to accept the task. Consequently, other tasks might not be processed anymore. These tasks have to be allocated again. In these processes a reallocate counter is increased. If the reallocate counter exceeds a predefined value, a new vehicle is created, all orders are returned to the task agent, and the procedure is started again.

---

[9]The benchmark instances are available at: http://www.sintef.no/projectweb/top/problems (cited: 1.9.15).

If all orders are successfully allocated to feasible tours, a post-optimization process is started. The vehicle agent removes a set of tasks from its tour, which has to be reallocated by the allocation agent. Consequently, the tours are further improved with respect to the objective function implemented in the decision-making process of the vehicle agents (in this case the objective function is to minimize the distances driven). The results show that their multiagent system computes adequate solutions. For instance, considering the primary optimization criteria of reducing the number of vehicles, the solver determines the best-known solution in 30.3% of all problem instances included in the VRP benchmarks of Solomon (1987) and Homberger and Gehring (2005) and in 11.7% of the problem instances included in the PDPTW benchmark on of Li and Lim (2001) (Kalina and Vokřínek, 2012b, p. 1563).

3.4.3 Conclusion and Discussion of Multiagent Systems in Transport Logistics

Table 3.3 gives an overview of the presented multiagent systems in transport logistics. It shows their most relevant aspects and allows comparing them in a structured way. Table 3.3 starts with approaches with a rather coarse-grained representation level. Fischer et al. (1996) were among the first to present a multiagent-based approach for the scheduling and planning of transport processes. Since then, several authors have adapted the general idea of Fischer et al. (1996) for the development of other multiagent systems. While the first approaches predominantly follow the coarse-grained modeling, recently developed systems implement rather fine-grained models, in which the decision-making is increasingly shifted from central agents to more decentralized agents which represent smaller logistic entities. This trend could well be related to technological advances such as CPS, which gives identities to smaller objects. Nevertheless, also in systems with a fine-grained multiagent-architecture centralized components play a major role. For example, order and vehicle agents are often applied to provide relevant information to a central managing agent, which is responsible for the decision-making, supervises the allocation, auctions orders to vehicle agents, and/or generates new agents. These systems are generally built around a central controlling agent and do not fully exploit the advantages of a decentralized multiagent system.

Table 3.3: Comparison of investigated multiagent systems in transport logistics.

Authors	RL[a]	CEN[b]	COM[c]	PNEG[d]	RA[e]	OPT[f]	SEQ[g]	PRE[h]	AAS[i]	Evaluation[j] B	S	A	cf. P.
Fischer et al. (1996)	C	CNT	ECNP+Other	yes	IA	yes	yes	no	no	yes	no	no	48
Bürckert et al. (2000)	C	CNT	ECNP+Other	yes	IA	yes	yes	no	no	no	no	no[k]	49
Thangiah et al. (2001)	C	CNT	CN	n.a.	IA	yes	yes	no	no	yes[l]	no	no	50
Perugini et al. (2003)	C	CNT	ECNP	yes	n.a.	no	yes	yes	yes	no	yes[m]	no	51
Dorer and Calisti (2005)	C	CNT	n.a.	no	IA	no	no	yes	no	no	yes	no[n]	52
van Lon et al. (2012)	C	DC	-	yes	GA	no	no	no	yes	no	no	no	53
Kohout and Erol (1999)	F	DC	CN	n.a.	IA	yes	yes	no	no	no	yes	no	54
Leong and Liu (2006)	F	CNT	Other	no.	IA	yes	yes	yes	no	yes	no	no	55
Mes et al. (2007)	F	DC	V+Other	yes	IA+BNB	(yes)	yes	yes	yes	no	(yes)	no	56
Barbucha and Jędrzejowicz (2008)	F	CNT	CN	no	IA	no	yes	yes	no	yes	no	no	57
Zhenggang et al. (2009)	F	CNT	CN	no	IA	no	yes	yes	no	yes	no	no	58
Himoff et al. (2005)	F	CNT	n.a.	n.a	n.a.	yes	n.a	n.a	no	no	no	yes[o]	59
Himoff et al. (2006)	F	CNT	n.a.	no	n.a.	yes	yes	no	no	no	no	no	59
Glaschenko et al. (2009)	F	DC	n.a.	yes	n.a	yes	no	yes	no	no	yes	no	59
Mähr et al. (2010)	F	DC	V+Other	yes	IA+Other	yes	no	no	yes	no	no	yes	59
Vokřínek et al. (2010)	F	CNT	CN	n.a.	IA	yes	yes	no	no	yes	no	no	63
Kalina and Vokřínek (2012b)	F	CNT	CN	n.a	IA	yes	yes	no	no	yes	no	no	66

a: Representation Level (RL): The approach has a rather coarse-granular (C) or a rather fine-granular (F) representation level.

b: The approach has essential centralized components (CNT) or it fully exploits the advantages of decentralized multiagent-based structure (DC).

(continued)

(continued)

c: Communication (COM): The applied communication protocol including variations of these (CN: Contract-Net (cf. Section 3.2.2); ECNP: Extended Contract-Net Protocol and variations (cf. Section 3.4.1); V: Vickrey auction (cf. Section 3.2.1)).

d: Parallel Negotiations (PNEG): Agent negotiations may be performed in parallel.

e: The routing algorithm which is applied by the agents (IA: Solomon's Insertion Algorithm; GA: Genetic Algorithm; BNB: Branch-and-Bound Algorithm).

f: Continuous Optimization (OPT): The approach continuously optimizes the solution quality (e.g., by an anytime behavior).

g: Sequence (SEQ): The solution quality depends on the sequence in which orders are processed.

h: Pre-processing (PRE): A Pre-processing is applied, e.g., to accelerate the search or improve the solution quality.

i: Agents Act Selfishly (AAS).

j: Evaluation: The approach is evaluated with benchmarks (B), by simulating real-world processes and data (S), and/or in application (A).

k: However, the approach was transformed to a commercial system called *TeleTruck*.

l: However, the authors do not indicate which benchmark is applied.

m: However, the authors evaluate the system only in a very small scenario with a single managing agent and three transport service providers.

n: However, the approach has influenced a commercial system developed by Whitestein Technologies (cf. http://whitestein.com (cited: 1.9.15)).

o: No result provided.

On the one hand, this limits the degree of parallelization and concurrency to the low number of decision-making agents in the approaches of Dorer and Calisti (2005), of Leong and Liu (2006), as well as of Barbucha and Jędrzejowicz (2008). On the other hand, it reduces the required communication and coordination effort of the system and allows the decision-maker to consider more available information of all agents. Multiagent systems, having neither a centralized component nor a central agent which controls or maintains the overall system, are fully self-organizational systems, which reach the global goal by emergent behavior of autonomously acting agents. Further authors, e.g., Thangiah et al. (2001), present a rather distributed multiagent-based implementation of a centralized algorithm.

In most approaches agents communicate and negotiate by the contract-net protocol (Smith, 1980). Only Máhr et al. (2010) apply the Vickrey auction (Vickrey, 1961) to ensure stability and to prevent selfishly acting vehicle agents from manipulating the outcome of an auction. If a service provider participates in multiple negotiations with several service consumers concurrently, the so-called Eager Bidding Problem (which is intensively investigated by Schillo et al. (2002)) has to be solved. However, only a few authors investigate or reveal how to solve this problem. The Eager Bidding Problem appears when the service provider offers the same proposal to multiple bidders. In this case, the first bidder could accept the order with the consequence that the other proposals are out-of-date and cannot be accepted anymore. This problem appears only in multiagent systems which support concurrent negotiations. To solve the problem, Fischer et al. (1996, pp. 10-11) developed the Extended Contract-Net Protocol (ECNP), which adapts the classical contract-net to handle temporal grants as well. Other mechanisms which match the approach of Fischer et al. (1996) are presented by Schillo et al. (2002).

In multiagent systems for transport logistics, decision-making agents have to value transport tasks for deliberation which task to accept and which to reject. This valuation must include numerous aspects such as the additional time and kilometers required to service the tasks, time windows, weight, restrictions of dangerous goods, limitations of different means of transport, etc. Finally, each valuation is based on the tour the vehicle has to drive. The computation of most efficient tours (the TSP with additional constraints; cf. Chapter 2) presents a major challenge, because all constraints must be considered in a very short time and routes must be computed or updated in each decision-making process. Depending on the scenario, the number of decisions made by software agents can easily exceed $100,000$ in a single run in real-world applications. With the exception of Máhr et al. (2010) and van Lon et al. (2012) the authors mentioned in Table 3.3 either apply Solomon's Insertion Algorithm (Solomon, 1987) (or a variation) or they conceal the details of the applied algorithm.

Moreover, each tour and route planning algorithm which sovles a TSP requires a distance matrix between the stops which must be visited. In general, a shortest-path algorithm computes distances in a real-world infrastructure, while in benchmark problems it is sufficient to determine the Euclidian distances. Although the computation of shortest routes quickly becomes a critical time factor in real-world applications, none of the authors deals with it or even mentions this problem. For instance, calculating the distance matrix to solve a TSP with only 20 jobs (40 pickup and delivery stops) requires already $40 \times 40 = 1600$ shortest-path search queries.

Table 3.3 also shows that some approaches do not use a continuous optimization strategy (OPT). Thus, they stop the optimization process and get stuck early in a local optimum if a valid solution is found (Barbucha and Jędrzejowicz, 2008; Zhenggang et al., 2009). Especially in dynamic environments, these approaches fail to maintain tours and preclude a reactive system behavior, which is generally one of the main reasons for applying a multiagent system. Indeed, the approach of van Lon et al. (2012) has a reactive behavior without any continuous improvement strategy, but the system is, therefore, no longer proactive as each single transport instruction is assigned to the vehicle only after it has finished the last one.

In most solutions, the solution quality depends on the sequence in which orders are allocated to agents. Thus, it is simple to determine which tour has to be adapted to transport an incoming order at the lowest cost (Barbucha and Jędrzejowicz, 2008; Zhenggang et al., 2009). However, the solution quality differs significantly if the allocation sequence changes. This is especially true for approaches which are based on the algorithm of Clarke and Wright (1964) (Thangiah et al., 2001).

Pre-processing can increase the solution quality and/or reduce the running time of a system. However, only four of the investigated approaches apply pre-processing. Although pre-processing is generally implemented by a centralized component, this does not necessarily limit the advantage of flexibility and robustness of a multiagent system. For instance, Glaschenko et al. (2009, p. 31) first split up the catchment area in several subareas to limit the communication to relevant agents, which are then responsible for objects located only in nearby districts.

The majority of the investigated systems are cooperative multiagent systems. Thus, the agents do not necessarily try to maximize their own profit and they have to reveal their plans and confidential data, e.g., to a central managing agent. New orders might also be accepted if the costs are increasing and higher than the revenue. In addition, the approaches, which apply the contract-net protocol are not stable so that cheating vehicle agents can manipulate the outcome of an auction by submitting deceptive valuations. As a result, Glaschenko et al. (2009, p. 33) implement additional surveillance techniques to identify cheating participants using information and communication technologies. Máhr et al. (2010) apply the Vickrey auction to ensure that bidders reveal their true valuations. Although in the multiagent system of Máhr et al. (2010, pp. 10-11) there is also a deficit of autonomy if containers are changed between vehicle agents, it generally consists of cooperating but selfishly acting agents.

The presented investigations substantiate the assumption of van Lon et al. (2012), that in multiagent-based approaches in transport logistics there is a lack of evaluation. A few systems are, e.g., evaluated on the well-known and often applied benchmark of Solomon (1987) or the benchmark of Homberger and Gehring (2005), which allow comparing the systems with each other. On the one hand, the benchmarks are primarily designed to compare Operations Research (OR) approaches solving static problems. Thus, static classical OR methods which are developed with the specific benchmark problem in mind outperform the multiagent systems, because they cannot profit from their extraordinarily flexible and robust behavior in dynamic environments. In addition, they do not benefit from the possibility of considering the individual requirements of heterogeneous goods and transport facilities. On the other hand, these benchmarks are nevertheless essential for measuring and comparing the solution quality of multiagent systems.

In order to investigate the system's flexible and reactive behavior in dynamic environments, Máhr et al. (2010, pp. 15-18) simulated scenarios based on real-world data from industrial partners. Similar to Mes et al. (2007) they prove that a multiagent system is competitive with or even outperforms the selected OR methods in dynamic environments. Moreover, Glaschenko et al. (2009) show in a case study that multiagent systems optimize dynamic real-world processes. Unfortunately, these systems are not evaluated on benchmarks. Thus, it does not become clear if the results profit from an *incomplete* modeling, inefficient real-world processes, or shortcomings of the OR method which the agent system is compared to. It is still an open question how efficient these computed solutions are compared to optimal or best-known solutions. Some approaches are also applied in real-world operations. Unfortunately, only rare performance measurements are provided which allow evaluating the success.

3.5 Summary and Conclusion

This chapter introduced multiagent technology, autonomous logistics systems, and existing multiagent systems, which have been developed to optimize processes in transport logistics. Especially in dynamic environments, multiagent systems are the technology of choice for the planning and scheduling of transport processes because of their reactive, proactive, and flexible behavior. Nevertheless, applying them to all kinds of transport problems is not necessarily suitable. In well-structured, static problems, in which all data and required information are centrally available, multiagent systems would only increase the complexity and communication effort. For instance, centralized Operations Research solvers or meta-heuristics are established approaches which have extensively been evaluated and applied for transport planning and scheduling (cf. Section 2).

However, most real-world transport processes need to consider domain-dependent requirements to create flexible and adaptive plans and to ensure a robust system behavior. Plans must then be extended immediately within a few seconds, because of new incoming orders or external events such as delays and other unforeseeable changes in the environment. Generally, there is not enough time to recompute the whole solution from scratch by an optimal solver. Even the recalculation by a fast heuristic may not be expedient because this might change parts of the tour which are not affected by the environmental change. In addition, a complete recalculation can lead to significantly changing tours in order to achieve marginal improvements. This confuses drivers, reduces the usability of the system, and in the worst-case it precludes the system from its introduction in real-world operations.

In multiagent systems, agents are *natural metaphors* for logistic objects (cf. Page 38). This significantly facilitates the modeling and implementation of dispatching systems which fulfill the individual demands of transport companies. In contrast to most traditional and generally centralized approaches, it is cost and time intensive to include company-specific requirements or to adapt a method to handle problems with slightly changing objective functions or constraints. The reason is that equations and heuristics are designed to solve either standard problems or problems with specific individual customer requirements. Their modification often results in the development of a completely new approach. An appropriate example is to support particular objective functions of the vehicles, which might consider individual parameters such as gas consumption, expert knowledge of the driver, or amortization of the vehicle. Moreover, the more constraints have to be considered the more complicated and less comprehensible equations become. This increases the susceptibility to errors and makes it difficult for domain experts and users, who generally have restricted knowledge on Operations Research and meta-heuristics, to understand and follow system assumptions and relations as well as the traceability of results. The goal is to keep the system as simple and extendable as possible and to enable operators to principally understand how solutions are computed and the way constraints are considered by the solver.

Finally, the multiagent-based structure ideally improves the monitoring and analysis of processes. Agents can provide full information of the represented object and report status data in real-time. This data is not only exploited to increase the efficiency, reliability, and robustness of logistic processes, but also to optimize dependent production processes. For example, operators can change the job schedule of a production plant at an early stage if delays are indicated in time.

As show in Section 3.4, there are several multiagent systems for planning and scheduling in transport logistics. However, these systems still have to many weaknesses to handle the future logistics challenges described in Section 1.1 in real-world applications. At first, some authors focus on the general approach and the design of the multiagent system's architecture, but fail to reveal relevant information, e.g., details about the negotiation, constraints, and the algorithms applied in the agent's decision-making processes (e.g. Himoff et al., 2005, 2006; Perugini et al., 2003). Therefore, it is hardly possible to evaluate their approaches in detail. Although it is one of the main advantages of decentralized systems, some approaches do not support concurrent negotiations, which is necessary to parallelize hardware intensive decision-making processes of agents in the multiagent system (Thangiah et al., 2001; Barbucha and Jędrzejowicz, 2008). Thus, only a few consider the Eager Bidder Problem (cf. Schillo et al., 2002), which arises in concurrent negotiations. Furthermore, some approaches are designed to solve static problems only. They do not implement any continuous improvement strategy (cf. Thangiah et al. (2001)), or they completely recompute a solution with an additional vehicle if the original amount of vehicles is not sufficient. This results in the reallocation of all orders and new tours of all vehicles.

In general, the implemented intelligence of the agents is restricted. With the exception of Mes et al. (2007) and van Lon et al. (2012) all authors apply Solomon's Insertion algorithm (Solomon, 1987) or a variation of it to solve a standard Traveling Salesman Problem (TSP) (with pickup and delivery constraints as well as with time windows in capacity constraints). Only Mes et al. (2007) applied an optimal branch-and-bound solver, which considers hard and soft constraints in the route planning. However, the agents only handle small problems: the problems they solve consider full-truckload orders only, in which each truck has to visit a maximum of 10 stops. Máhr et al. (2010) further extended the insertion by substitutions which apply the knowledge gained in the Vickrey auction. In addition, the problem to determine shortest paths between stops is excluded in all investigations, although it is highly relevant especially for real-world applications in dynamic environments. In static problems all relevant stops are known and distances and the estimated driving time are not changing. Thus, shortest paths could be computed in advance. The importance of high-performance shortest-path computation is growing in dynamic cases, because traffic congestion, changing order situations with new incoming orders at previously not considered locations, and other external events influence the shortest-path computation.

One explanation for not intensively focusing on high performance decision-making processes could be that the number of negotiations and decision-making processes to improve solutions seems to be relatively small in the scenarios investigated by the authors. For instance, Máhr et al. (2010, p. 108) start only a single reallocation (or improvement) negotiation per hour and process only 64 orders per scenario. Other authors conceal the amount of improvement processes. It is obvious that such a low number of (concurrently) running negotiations also reduces the probability of (or even totally avoids) the Eager Bidding Problem and significantly simplifies the coordination and synchronization of communicating agents. Other authors, such as Mes et al. (2007), only evaluate their approach in small scenarios in which high performance decision-making is unnecessary, because the routing problems contain less than 10 orders. These small problems can be solved optimally within an adequate period of time. Similarly, Perugini et al. (2003, p. 24) modeled a scenario with a single managing agent and three transport service providers only.

Finally, there is a lack of evaluation to judge the quality, effectiveness, and behavior of developed approaches (especially in dynamic environments). Some systems have been applied in real-world operations, others have outperformed selected Operations Research approaches in dynamic environments, some are evaluated with static benchmarks, while further authors show by case studies with an industrial partner, that their multiagent-based approach significantly increases the efficiency of processes. However, not one approach has been extensively evaluated by benchmarks plus simulations based on real-world data plus practical application in real-world operation.

4 The dispAgent Approach

This chapter presents the dipsAgent approach which optimizes the planning and control processes in transport logistics and allows handling domain-specific demands in complex and dynamic environments. The challenge is to preserve, extend, and optimize suitable concepts, components, and approaches in order to overcome the weaknesses described in Chapters 2–3 and to use multiagent-based autonomous processes in real-world applications.

To this aim, the level of concurrency must be increased and the intelligence of the agents must be improved in such a way that all available information and the agents' objectives are directly considered in their decision-making by tour planning algorithms. This also includes a detailed investigation of the shortest-path problem which is highly relevant for real-world operations. Moreover, a continuous improvement strategy is ideally applied to increase the solution quality. Since it is required in some domains, the strategy should have an anytime behavior to stop the system if the result is required by the operator. Therefore, the multiagent system should monotonically increase the solution quality of computed tours after a valid solution has been found.

Firstly, Section 4.1 provides a general overview of the dispAgent approach. Secondly, Section 4.2 presents the implemented pre-processing algorithm. The centralized static analysis is a rather rough planning, which computes a generic static solution to improve the runtime performance. The multiagent system improves the generic tours, considers domain-specific requirements, and ensures a flexible, reactive, and robust behavior in real-world dynamic environments. Section 4.3 describes stable communication and negotiation mechanisms which allow for a high level of concurrent computations. Next, Section 4.4

presents high performance algorithms which the agents use for decision-making. The section further outlines how the quality of the agents' decision-making is increased by considering available information and the agents' objectives directly in the tour planning algorithms. Section 4.5 focuses on efficient and parallel computations of shortest paths. Finally, Section 4.6 concludes this chapter.

4.1 An Overview of the dispAgent Multiagent System

Similar to approaches with a fine-grained multiagent model, agents represent orders and vehicles. With the exception of the rough planning applied as pre-processing, centralized components are avoided so as to fully exploit the advantages of the decentralized multiagent structure, which include, e.g., improved robustness and an increased degree of parallelization.[1] The order agents start auction-like negotiations with available vehicle agents to find a proper transport facility. The vehicle agents value received *call-for-proposal* messages and send an offer or refuse a transport. Within their decision-making process, the applied route planning algorithms directly consider the individual properties of the vehicles, the orders' requirements, as well as the overall order situation by processing the information revealed in the Vickrey auction (cf. Section 3.2.1). In negotiations, already accepted orders might not be transported anymore if another order has a higher priority. The agents of these postponed orders must then start new negotiations with other vehicle agents. Concurrently, order agents which have already found an appropriate transport facility, continue negotiating with the vehicle agents to check if there is another transport service provider which fulfills the orders' demands at lower costs. In addition, these *dynamic negotiations* enable fast reaction to changing environmental influences and to varying order situations.

---

[1]Even if the centralized pre-processing component is deactivated or fails, the system remains fully operational without it.

On the one hand, parallel negotiations require synchronization mechanisms to ensure the consistency of the system. For instance, a vehicle agent often sends proposals to several order agents in parallel. Then, the tour of the vehicle changes when the first proposal is accepted by an order agent and other proposals might be out of date (cf. the Eager Bidding Problem described by Schillo et al., 2002). On the other hand, the system profits from distributed concurrent computations, which can also be performed on high performance computers, e.g., in the *cloud*.

The dispAgent software implementation uses the FIPA-compliant Java Agent Development (JADE) framework (Bellifemine, Caire, and Greenwood, 2007) as the underlying agent management platform. An agent management platform is responsible for the message transport and has a *directory facilitator* (agent) which provides yellow pages services for agents. In addition, JADE includes abstract agent behaviors which can be extended to implement, e.g., *parallel behaviors*, *finite-state-machine behaviors* (FSM behaviors), or *cyclic behaviors*. In JADE, each agent has its own process in the operating system. As a result, as many agents as there are cores available can physically run at the same time.

The implemented multiagent system (MAS) is also extended by further synchronization mechanisms which allow the evaluation and simulation of dynamic transport scenarios within the *Platform for Simulation of Multiple Agents* (PlaSMA) (Warden, Porzel, Gehrke, Herzog, Langer, and Malaka, 2010) (more details about PlaSMA are provided in Section 6.4.1). In dynamic simulations, there exist different notions of time. "*Physical time* refers to time in the physical system [, while] *simulation time* is an abstraction used by the simulation to model physical time"(Fujimoto, 2000, p. 27). In real-world operations, synchronization is based on the physical time only. However, in simulations the simulation time must be considered as well, e.g., to avoid that agents receive messages which are sent in the future (wrt. the simulation time and not to the physical time). Thus, all agents must additionally synchronize their interaction with the environment (the modeled world) as well as the message exchange with other agents to ensure correct, consistent, and reproducible simulations. Therefore, a general generic time management service for distributed multiagent systems is developed by, e.g., Braubach, Pokahr, Lamersdorf, Krempels, and Woelk (2006) and Pawlaszczyk and Timm (2007). The synchronization mechanisms of the PlaSMA framework and their usage is presented in detail by Gehrke, Schuldt, and Werner (2008).

4.2 Rough Planning

In order to reduce the runtime of the system, a static pre-processing can be applied. In real-world scenarios, multiple orders have already been commissioned before the operational processes start, e.g., in the morning. Without any pre-calculation, all order agents start negotiations in parallel when the planning is started. For example, if 1,000 shipments have to be allocated to 50 available vehicles, at least 50,000 negotiations are started and the vehicle agents perform their cost-intensive decision-making processes to value each order. However, these high amount of parallel negotiations might cause conflicts and reallocations, because the probability increases that proposals are *out-of-date*. In addition, order agents often decide to switch the transport provider, because the vehicle's tours change significantly in the initial tour construction phase.

In order to reduce this computational overhead, all available order and vehicle agents send their transport requests as well as their transport resources and capabilities to a cluster agent. The cluster agent collects all messages and computes an initial (possibly constraint-violating) allocation. Next, it sends the allocated orders to the vehicle agents. Each vehicle agent computes a constraint-satisfying tour which ideally contains all assigned orders. Therefore, it starts its individual decision-making process which considers all domain-dependent and customized requirements. The agents of orders which can be transported by a vehicle receive a *transport-is-possible* message from the respective vehicle agent and start new negotiations to check if there is an even more suitable transport option. The agents of orders which remain unserved receive a *transport-not-possible* message and start new negotiations with other vehicles by themselves. Consequently, the centralized cluster agent is removed and the agent system is fully decentralized from this point of the operation. Nevertheless, the autonomy of the agents is not affected at any time. For example, if an agent is unsatisfied with an assignment, it can look for another vehicle. If the centralized cluster agent is not available, the agents start direct negotiations.

In principal, any static and even centralized approach could be applied as pre-processing as long as it covers the specific requirements of the problem. For instance, a solver for the static Vehicle Routing Problem (VRP) or Pickup and Delivery Problem (PDP) might be applied (cf. Chapter 2). However, also the static problems are NP-hard and require computationally intensive approximation algorithms to be solved at all.

Therefore, a cluster algorithm is used which assigns available orders to several sub-areas. There should be as many sub-areas as there are vehicles. Thus, the exact number of clusters is known. In addition, particularly in real-world problems, nearby orders are mostly processed by the same vehicle to reduce the overall distances driven. As a result, a partitional k-means-clustering (MacQueen et al., 1967) is implemented to quickly determine a suitable allocation. As each order has a pickup and a delivery location, the central point between the pickup and the delivery location is the relevant coordinate to determinate the cluster-affiliation and to update the cluster center. Consequently, this approach additionally allows for solving problems with central depots, which are modeled as the identical pickup or delivery location of orders. To compute the distance between two coordinates C_1 and C_2, the distance on a sphere is computed by the *haversine* formula (Sinnott, 1984, p. 159).

Definition 4.1 (Sphere Distance). *Let lat_1 and lat_2 denote the latitude in degree of coordinates C_1 and C_2 respectively. Let lon_1 and lon_2 denote the longitude in degree of coordinates C_1 and C_2 respect-*

ively. Let the function toRadian(x) transform an angle x from degree in radians and R be the earth's radius in Northern Europe which is approximated by 6371 (km). Then, the distance d between C_1 and C_2 on a square is defined by

$$\Delta lat = lat_2 - lat_1 \tag{4.1}$$

$$\Delta lon = lon_2 - lon_1 \tag{4.2}$$

$$a = (\sin(\Delta lat/2))^2 + \cos(lat_1) * \cos(lat_2) * (\sin(\Delta lon/2))^2 \tag{4.3}$$

$$c = 2 * atan2(\sqrt{a}, \sqrt{1-a}) \tag{4.4}$$

$$d = R * c. \tag{4.5}$$

4.3 Concurrent Continuous Multiagent Negotiations

In the negotiations each order agent looks for the most appropriate vehicle by a reverse Vickrey auction (cf. Section 3.2.1). Therefore, vehicle agents take on the role of participants, while the order agent initiates the negotiation. The vehicle agents concurrently apply routing algorithms to determine their costs for transporting additional good(s). They send their results back to the order agent which accepts the offer with the least costs. Next, the order agent informs the winning vehicle about the acceptance of the proposal and reveals the second best offer. Similar to Máhr et al. (2010, p. 9), the winning vehicle agent considers this information in its future decision-making. The amount of the offer indicates if the vehicle agent with the second best offer would skip another transport request in order to handle the new order. In this case, the priority to transport the order increases for the winning agent, because other vehicles do not have sufficient capacities to handle the transport request. When the vehicle agent computes further proposals, this increased priority of the order is directly considered by the tour-planning algorithms. As a result, the vehicle's valuation depends on its own tour as well as on the available resources of the other vehicles.

As a result, it is possible that already accepted orders are not processed anymore, because an order with a higher priority is transported instead. Then, not-processed orders receive a *transport-not-possible* message and start a new negotiation with the other vehicle agents to look for alternative transport service providers. In order to avoid loops of recurring substitutions of orders, the order agents can also accept the second best proposal instead of the best one with a certain probability. This process is continued until all orders are allocated to vehicles (or allocation has at least been attempted, in case it is not possible to service the order).

As the vehicle agent participates in several auctions concurrently, the resulting Eager Bidding Problem has to be solved. For instance, if the vehicle's order situation is changed between the time the proposal is offered and the time of acceptance, the offered proposal is out of date and has to be updated. To identify this case, each vehicle agent maintains a *hash map* with all proposals as well as the creation date of the plan which was active when the proposal was computed. If the creation date of the current plans differs from the date saved in the *hash map*, the plan has changed. Thus, a recalculation of the proposal is required. However, the proposal is only accepted if no other order is substituted. Otherwise, the vehicle sends a failure message and the order agent has to start a new reverse Vickrey auction.

4.3.1 The Communication Protocol for Dynamic Negotiations

Unfortunately, if only these initial negotiations would be applied, the quality of the results would depend on the randomly chosen sequence in which orders are allocated to or accepted by the vehicles. To counter this effect, the dispAgent approach allows for continuous concurrent negotiations between agents to identify more suitable and cheaper transport options, which might also result, e.g., from a changing order situation or other external events.

While the extension and integration of the agent behaviors for sequential allocations by the contract-net protocol (or any other type of auction) is comparatively straightforward, it is more complex to reduce conflicts, to synchronize negotiations, and to ensure a persistence system state in these continuous and highly concurrent communications. Other authors avoid this problem by permitting an agent to participate in several negotiations at the same time or by setting the status of a processed order on hold. For instance, Máhr et al. (2010) minimize the possibility of conflicts by the introduction of a randomized waiting behavior. The maximum waiting time is limited by a parameter. In their simulated scenario, which is comparatively small with merely 65 orders, they set this parameter to one hour. Thus, they restrict the number of negotiations to approximately one per minute. Therefore, this strategy significantly limits the degree of parallelization (and even avoids concurrent negotiations), increases the runtime, and has a negative impact on the solution quality. Alternatively, the order agents could participate in several negotiations concurrently and benefit from a higher level of parallelization, because in most cases the negotiations do not interfere. It is obvious that this improves the result quality.

Nevertheless, in the dispAgent approach a randomized waiting behavior is adapted and integrated, because the possibility to create a conflict is increasing when an order agent starts a new negotiation immediately after the last one. Therefore, it is avoided that simultaneously created agents start their negotiations exactly at the same time.

Furthermore, it is unlikely that a more suitable transport facility will be found some milliseconds later, because there is not enough time for changing the allocation. In the dispAgent implementation, the minimum waiting time is limited to a small parameter value (e.g., one second). A suitable value of this parameter depends on the degree of parallelization and on the runtime of the decision-making of a vehicle agent, which is influenced by the used hardware.

In order to reduce the number of redundant negotiations, in the approach of Máhr et al. (2010, p. 7) the order agents start a new auction only when the plan of any truck has been changed. However, this restricts the autonomy of the order agents, because they rely on the behaviors of vehicle agents. Moreover, it precludes the order from reacting to internal changes, e.g., spontaneously varying time windows or pickup and delivery locations. Thus, the dispAgent approach follows the strategy that orders continuously observe the environment and check if suitable changes are possible or even necessary.

A general difference to other approaches, which apply a multiagent-based implementation of the k-opt improvement strategy of Lin (1965b) (such as Máhr et al., 2010), is that the decision to change orders between vehicle agents is not made by the vehicle agents but instead by the order agents themselves. Thus, the autonomy of order agents is guaranteed and the decision-making authority is not shifted to the vehicle agents. Consequently, the order agents decide themselves which part of information should be revealed and which transport provider is selected. For instance, in real-world scenarios, only selected drivers have permission to enter the premises or are allowed to transport dangerous goods. Following the dispAgent approach, these constraints can simply be considered by the order agents themselves. In addition, this example shows that the modeling ensures that confidential data is kept private and only revealed to selected drivers and that the multiagent system remains a fully self-organized system.

During operations, the transportation costs for an order are changing. For instance, in case of an unexpected event such as traffic congestion, the real costs might increase. Thus, the order agent must

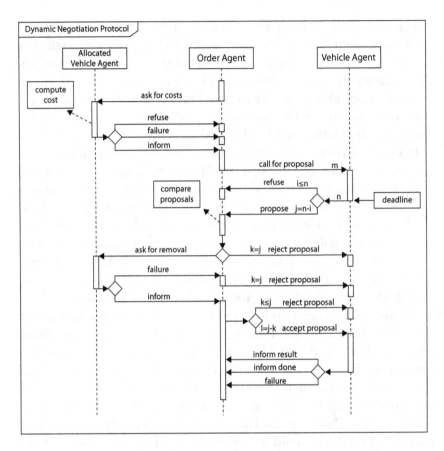

Figure 4.1: The dynamic negotiation protocol which is applied to improve the tours continuously.

be aware of its current estimated costs before starting a negotiation with another transport facility.

Figure 4.1 shows the protocol which is developed for dynamic concurrent negotiations. The protocol is initiated by the order agent, whose order is already allocated to a vehicle. The order agent asks the responsible vehicle agent for its current costs to transport the order. To compute the costs the vehicle starts its decision-making process. If the order cannot be handed out, because the transport is already in progress or the order is already loaded, the vehicle agent sends a *refuse* or a *failure* message. Otherwise, it determines the costs and sends it back to the order agent. Next, the order agent starts a new reverse Vickrey auction with other vehicle agents. The main difference of this part of the protocol to the standard reverse Vickrey auction is that the order agent accepts a proposal only if the offered price is lower than its current transport costs and the order was successfully removed from the tour of the former vehicle. If the order agent prefers to conceal its costs, it decides later which proposal is accepted or rejected. However, revealing the costs reduces the communication effort significantly, because the vehicle agents can send a *refuse* message and step out of the negotiation. If any *failure* message is sent, the negotiation is aborted and started again later. In the unlikely case the order has already been removed from the tour of its former vehicle, the respective order agent starts a reverse Vickrey auction to find a new transport service provider.

On the one hand, this continuous improvement strategy has several advantages. Firstly, the decision-making competence remains with the order agents, which ensures the autonomy of the agents. Secondly, an allocation is monotonically improved in the negotiations.[2] This is an essential feature for real-world applications, as this allows stopping the system or a negotiation at an arbitrary point in the process and return the best result found, e.g., when the tour plan is required by the operator or driver. On the other hand, in contrast to k-opt improvement techniques, the transfer is limited to single orders. Thus, correlated changes of multiple orders are not considered. However, even a k-opt strategy would not guarantee to find the optimal solution and is incomplete from an algorithmic point of view.

---

[2]With the exception of the unlikely case that a negotiation is canceled after the order has been removed from the former vehicle.

4.3.2 Stopping the MAS in Consistent States

In general, the MAS is designed to run continuously. Nevertheless, it is possible to stop the system at any point during operations, e.g., to analyze the tours or because of hardware restrictions if the system is running in the cloud. Therefore, a termination agent is created which sends all available agents a message with a termination request. The general idea is that the agents receive all messages, stop all their behaviors, send their results back to the termination agent, and finally finish (or stop) their *lifetime*. The result they send includes thier current tour and some meta-data. The result which an order agent sends indicates if the order is being transported or still unserved.

However, there are several issues that have to be resolved when this termination mechanism is implemented in a highly parallelized MAS. On the one hand, the whole stopping process could require an unacceptable amount of time if the agents processed all message still included in their message queue (inbox) and, thus, finished all their negotiations first. On the other hand, it must be ensured that the MAS is terminated in a consistent state if negotiations are aborted. The agents perform several behaviors in parallel. For instance, a complex FSM behavior is responsible for the dynamic negotiations and another cyclic behavior is waiting for an incoming message such as a *transport-not-possible* message. Although the agent processes its behaviors by applying a round-robin scheduling, it is hard to predict which behavior is scheduled next, because the negotiation protocols, e.g., the contract-net protocol (cf. Section 3.2.1), consist of multiple *atomic* behaviors that are consolidated in a complex behavior. The execution of the next *atomic* behavior often depends on the content of the message which is received or processed. Now, it is assumed that a vehicle agent stops its behaviors although it has received a *call-for-proposal* message, e.g., in the negotiation in Figure 4.1. In this case, the order agent could wait an infinitely long time but would never receive a proposal. To avoid this deadlock if no vehicles are available anymore, the order agent is allowed to finish its negotiations and terminate itself. Another parallel running behavior checks this condition by frequently asking the *directory facilitator* about the availability of vehicle agents. However, in rare cases this could result in an inconsistent state of the MAS, because agents have no possibility to ensure and check that a sent message is received by the sender. Such a case can be retraced by Figure 4.1. Assuming that the order agent sends an *ask-for-removal* message, the receiving vehicle agent receives and processes the message, removes the order from its tour, and sends an *inform* message back to the order agent. Immediately after this behavior, the vehicle agent terminates. Then, the order agent could check if the vehicle agent is dead, and terminates itself before it had received the *inform* message from the agent management system (AMS), which is responsible for transferring messages. Thus, the status of the order remains *allocated* and the transmitted result is inconsistent. Consequently, it is essential to wait for a sufficient length of time to receive all important messages from the AMS before the order agent terminates in order to process all relevant messages (irrelevant messages could be removed from the inbox to save time). An adequate waiting time depends on the amount of concurrently running negotiations and used hardware resources.

However, it is possible that an order agent receives a *failure* message during a dynamic negotiation, is unserved, and the system terminates (although there might be another vehicle which has sufficient resource capacities to transport the order). To avoid this, the vehicles continue to process messages of unserved orders for a certain amount of time even after they received their termination request. In ongoing negotiations, the vehicles only accept proposals if no other orders are substituted instead.

In conclusion, the mechanism delays termination to ensure that enough time remains to receive and process all relevant messages in these highly parallel negotiations. This is essential for guaranteeing that the returned result is consistent and no orders are missing or considered twice.

4.4 The Agents' Decision-Making

The agents' decision-making is the most cost- and memory-intensive operation of the MAS with high computational demands. During negotiations, vehicle agents must compute proposals, deliberate which service request they want to accept, and estimate the required effort for transporting the goods. These valuations are mainly based on the additional distance or time which the vehicle requires to service a task. The problem refers to a generalization of the NP-hard Traveling Salesman Problem (TSP) (Christofides, 1976), in which several other constraints must be considered such as service time windows, limited vehicle capacities, and also handling and service times. Beside the individual preferences of the vehicle agent, the planning algorithm directly considers the additional knowledge about environment and current order situation in order to increase the solution quality and performance. The vehicle is aware of the second best proposal, which reveals whether or not there is another vehicle with sufficient capacities to transport the order. It processes this information by assigning different priorities to the orders. The solver considers these priorities during the search process in order to guide the search into promising directions.

In contrast to a standard TSP solver, the optimization objective is different. The highest prioritized goal is to include as many orders as possible if it is not possible to transport all orders. This goal is further subdivided. Orders with higher priority than conventional orders have so-called premium stops which must be visited with higher priority than the stops of conventional orders.

Definition 4.2 (Premium Stop). *Let premium : $S \to \{0, 1\}$ denote a function which defines if a stop $s \in S$ is a premium stop by*

$$premium(s) = \begin{cases} 1 & \textit{if } s \textit{ is a premium stop} \\ 0 & \textit{otherwise.} \end{cases} \tag{4.6}$$

As a result, the classical objective function to find a feasible solution which contains all stops (cf. Eq. 2.3-2.6) and minimizes the total costs (cf. Eq. 2.2) must be adapted. The goal is to find a tour that maximizes the number of premium services with highest priority

$$max \sum_{i \in S} \sum_{j \in S} premium(i) \cdot x_{i,j} \qquad (4.7)$$

and conventional orders with second-highest priority

$$max \sum_{i \in S} \sum_{j \in S} \neg premium(i) \cdot x_{i,j}. \qquad (4.8)$$

With third-highest priority total costs must be minimized (cf. Eq. 2.2).

Below, Section 4.4.1 presents an optimal time- and memory-efficient algorithm for smaller problems. Section 4.4.2 examines a novel approach for solving large-size problems, which computes state-of-the-art solutions within a short time. Note that both solvers must fulfill exactly the same objective functions and constraints if they are applied by the same agent for solving small and large-size problems. Otherwise, the solvers might optimize the agent's objectives in the opposite direction.

4.4.1 Optimal Decision-Making

In order to optimally solve small and constraint problems, the agents apply a Depth-First Branch-and-Bound (DFBnB) algorithm. Let N denote the set of all nodes (states) of the search tree. Using a standard DFBnB algorithm (Zhang and Korf, 1995), the solution quality improves continuously together with the global upper bound $U : N \rightarrow \mathbb{R}$. A lower bound can be computed by a function $L : N \rightarrow \mathbb{R}$. If the lower bound $L(n) = U(n)$ with $n \in N$, the search terminates and an optimal solution has been found. In most problems, additional pruning rules are applied to further cut subtrees and to reduce the search space if a constraint is violated. In general, the lower bound $L(n)$ is computed by an admissible heuristic $heuristic : N \rightarrow \mathbb{R}$ and by the current cost $g : N \rightarrow \mathbb{R}$ at node n with $L(n) = g(n) + heuristic(n)$. For solving TSPs there are multiple heuristics for the computation of adequate lower bounds. For instance, the (trivial) lower-bound heuristic $heuristic_{trivial}$ depends on the distance from the current state back to the depot. Obviously, $heuristic_{trivial}$ can be computed in $O(1)$ time. Since the *Assignment Problem* (AP) (Kuhn, 1955; Munkres, 1957) is a relaxation of the asymmetric TSP it can also be used as lower bound.[3] The AP can, e.g., be solved by the *Hungarian Algorithm* of Jonker and Volgenant (1986) incrementally in $O(n^2)$. Upper bounds can, e.g., be computed by applying the Karp-Steel patching (Karp and Steele, 1985).

However, to satisfy the initially mentioned hierarchical objective function, it must be avoided that the search tree is pruned by these general lower and upper bounds before all stops are included within a solution. The reason is that the heuristics neglect the prioritized objective functions (cf. Eq. 4.7 and Eq. 4.7). Only if the search reaches a leaf of the search tree in the maximum depth, established lower and upper bound pruning rules can be applied to accelerate the search and to minimize the total distance (cf. Eq. 2.2). Consequently, applying additional time-efficient pruning rules is essential. In addition, the algorithm described in this section ensures that it terminates when a fixed number of expansions is exceeded. As a result, the algorithm has an *anytime behavior* and finds better solutions the more time it keeps running (as long as the optimal solution is not reached). It returns the best-found valid solution if it is interrupted or the optimal solution otherwise.

An Optimal DFBnB Algorithm for TSPs with Priorities and Additional Constraints

Algorithm 1 shows the pseudo-code of the iterative implementation of the proposed DFBnB algorithm. Optionally, it is also possible to implement the algorithm with a recursive function. However, the

---

[3]In problems in which it is necessary to return to the depot, the distance matrix must be extended with by distance from the depot to the current node if the central depot is not the starting point. After solving the AP, the distance is subtracted to determine the current lower bound.

Algorithm 1 DFBnB for TSPs with Premium Service and additional Constraints

---

1: **procedure** CONSTRAINT-TSP-DFBnB(S,s_{depot},$maxExp$)
2:    $u \leftarrow estimateUpperBound()$
3:    $exp \leftarrow d \leftarrow 0$
4:    $tour \leftarrow best \leftarrow$ null
5:    $allIncluded \leftarrow false$
6:    $newNode \leftarrow$ CREATENEWNODE(s_{depot})
7:    $stack.push(newNode)$
8:    $N \leftarrow |S|$
9:    **while** $stack.size > 0$ **do**
10:       exp++
11:       **if** $maxExp \geq exp$ **then**
12:          **break**
13:       **end if**
14:       $currentNode \leftarrow stack.pop$
15:       $tour[currentNode.depth] \leftarrow currentNode.city$
16:       **if** COST($best$) > COST($tour$) **then**
17:          $best \leftarrow tour$
18:       **end if**
19:       **if** $d = N - 1$ **then**
20:          **if** $g+$ DIST($currentNode.city, depot$) < u **then**
21:             $u \leftarrow newNode.g+$ DIST($currentNode.city, depot$)
22:             $allIncluded \leftarrow true$
23:          **end if**
24:       **else**
25:          **for all** $s \in S$ **do**
26:             OPT: CHOOSE NEXT CHECKED NODE BY HEURISTIC
27:             $newNode \leftarrow$ CREATENEWNODE(s)
28:             **if** CHECKCONSTRAINTS($currentNode, newNode$) **then**
29:                **if** $allIncluded = false$ **then**
30:                   $stack.push(newNode)$
31:                **else if** $newNode.g+$ HEURISTIC($newNode$) < u **then**
32:                   $u \leftarrow newNode.g+$ HEURISTIC($newNode$)
33:                   OPT: UPDATE UPPER BOUND BY HEURISTIC
34:                   $stack.push(newNode)$
35:                **end if**
36:             **end if**

37:	**end for**
38:	**end if**
39:	**end while**
40:	**return** *best*
41:	**end procedure**

---

iterative approach can reduce the memory consumption, because it
is not necessary to save the *recursion stack*. The input parameters
are stops S, the start and end stop s_{depot}, as well as the maximum
number of expansions ($maxExp$). If the search should be complete
and optimal $maxExp$ must be set to infinity. In line 2 the upper
bound u is set to some reasonable estimate (it could be obtained by
using some heuristics; the lower it is, the more can be pruned of the
search tree, but in case no upper bound is known, it is set to infinity).
The initialized variables and constants in lines 2-8 are maintained
globally. The global variable *best* keeps track of the current best tour
(solution path). If a tour with lower costs is found, this tour is updated
as the best found result. The *stack* contains a set of nodes of the
search tree. Each node has several attributes such as the represented
city, the depth of the node in the search tree, the current capacity
the vehicle has at this node, the time which has currently elapsed at
this node, the current cost of the tour, and a heuristic value which is
the estimated time to return to the depot. Moreover, it maintains a
variable that indicates the visited cities.

The loop starting in line 9 either stops if the algorithm finds the
optimal solution (the *stack* is empty) or if it exceeds a maximum
number of expansions (line 11-13). Let $S' = \{s_0, \ldots, s_n\} \subseteq S$ with
$0 \leq n = |S'| \leq |S|$ denote a partial tour. The current city represented
by the node is added to the current tour at the position which is
equal to the depth of the search tree.[4] Whenever the tour is changed,
the algorithm checks by a cost function $cost : S' \to \mathbb{R}$ if a new
tour's valuation outperforms the current best tour and updates the

---

[4]It is assumed that this is possible at least for the first city.

variable *best* if necessary. In contrast to the costs determined by the transition between stops in the classical TSP (cf. Definition 2.1 in Section 2.1), the cost function also allows to consider dependencies and synergies between all cities of the tour. Therefore, the implemented cost function considers the priorities of premium services, e.g., by an extraordinarily high value such as 10^6. It is obvious that an increasing depth leads to a rising number of included orders. If all orders are included (lines 19-23), the third objective function which minimizes the total time (or distance) has to be fulfilled. For this purpose, the overall distance is saved as upper bound and further pruning rules can be applied to accelerate the search. As mentioned above, upper and lower bound heuristics are only activated if the best found solution includes all stops, because a longer tour including more stops is preferred. Nevertheless, the applied heuristics must be compatible with the COST function applied in line 16 to ensure that the algorithm is optimal. In case the search is continued (lines 25-34), the algorithm checks for all successors whether they have already been visited and included in the tour and whether they fulfill other constraints such as time windows and capacities. Depending on the constraints of the problem, the constraint check significantly reduces the search space. Only if this check is positive, a new node is created and added to the *stack*. Otherwise, the search tree is pruned at this node. Moreover, the search can be accelerated by a further heuristic, which determines the next successor stop that will be checked. Good solutions are found faster and leaf nodes reached as early as possible by checking promising cities first and guiding the search. For this purpose, the heuristic described in Section 4.4.2 which initializes the policy for the Nested Rollout Policy Adaptation (NRPA) algorithm can be applied (cf. Page 113). If the algorithm terminates before the maximum number of expansions is reached, it returns the optimal solution with regard to Eq. 4.7, Eq. 4.8, and Eq. 2.2. In case the vehicle does not return to the depot, this variation can easily be considered by changing line 20 (but note that the applied heuristic function in line 31-32 might be changed as well).

Optimal Time- and Memory-Efficient Constraint Checks

Beside heuristics, the CHECKCONSTRAINTS function is the most cost-intensive operation of the algorithm. Due to the fact that the changed objective function permits using the established upper and lower bound heuristics as long as the tour is not complete, the significance of a time- and memory-efficient implementation of the CHECKCONSTRAINTS function is even growing. Therefore, the CHECKCONSTRAINTS operation is implemented by efficient bit-vector comparisons in $O(1)$ constant time and space which allow checking millions of nodes in a few milliseconds.

Theorem 4.1. *Checking constraints in Algorithm 1 (line 28) for each search node for solving a TSP having either pickups or deliveries is done in $O(1)$ constant time and without a dynamic allocation of any additional space.*

Proof. At each node the visited stops, the current time, premium service information, and the current capacity of the tour are saved as a computer word w. It is assumed that the standard word length of, e.g., 64 bit, is sufficient for a suitable granularity to represent the constraints and the maximum length of a tour. Then the memory consumption on each node is bounded by $O(1)$.[5] All bit-vector, arithmetic, relational, and equality operations run in $O(1)$ in the standard assumption in the RAM model.[6] Due to the fact that all relevant information is saved at the node, no backtracking is required and no additional memory must be allocated for constraint checks. For instance, the current time as well as the latest possible time to visit the stop is saved at the node. Thus, the time constraints are checked at the node by

---

[5]If $|S|$ denotes the length of the tour which includes all stops (also the start and end point) and w the length of the hardware dependent computer word, then the memory consumption would be bounded by $O(\frac{|S|}{w})$ which can be assumed to be constant.

[6]If $|S|$ denotes the length of the tour which includes all stops (also the start and end point) and w the length of the hardware dependent computer word, then it is assumed that the complexity is of $O(\frac{|S|}{w}) = O(1)$.

`currentTime < latestTime`.

In pure pickup or delivery problems checking the capacity is done in a similar way by comparing the sum of all transported shipments to the maximum capacity of the vehicle. Moreover, if a stop has already been visited the check is done by bit-vector operations. For instance, if stop s, which is decoded as bitvector x, is visited can be checked by

`(((visited >> x) & 1L) > 0)`.

As a result, all constraint checks are performed in constant $O(1)$ time and space. □

However, this does not hold for the mixture of pickups and deliveries in a single problem because capacities varying. Delivered shipments release capacities for picking up more freight. In the ongoing tour, the released capacities have to be considered to ensure optimality. Therefore, two more variables for the capacity of all deliveries and the maximum capacity reached on the tour have to be maintained at each node.

Theorem 4.2. *Checking the capacity constraints in Algorithm 1 (line 28) for each search node to solve a TSP with simultaneous pickups and deliveries is done in $O(1)$ constant time and space.*

Proof. Let χ_M denote the maximum capacity at each node that the truck has reached on the tour and CC the current capacity of the truck. Moreover, let ω denote the weight of the order at stop s. Then, at each node χ_M is updated with

$$\chi_M = \begin{cases} \max(CC, \chi_M + \omega), & \text{if } s \text{ is a delivery stop} \\ \max(CC, \chi_M), & \text{otherwise.} \end{cases} \tag{4.9}$$

If M denotes the maximum capacity of the truck, the capacity constraints for adding a new order are satisfied by checking

$$M \geq \chi_M. \tag{4.10}$$

Consequently, all operations can be implemented by bit comparison. No backtracking is necessary to avoid the overloading of trucks on predecessor nodes by adding new delivery stops to the tour. □

Table 4.1 gives an example of these constraint checks in constant time and without the allocation of additional space. In this example, it is assumed that the maximum capacity of the truck is 4 units and the weight of each shipment is 1 unit. To avoid backtracking, CC and χ_M is maintained at each node. If the vehicle picks up a shipment, the current capacity is increasing. Adding a delivery stop to the plan does not affect the current capacity because loading the shipment was not considered up to this point. Nevertheless, the vehicle has to load the shipment at the depot before starting the tour. Therefore, χ_M is increasing. Consequently, it is not possible to add a delivery stop in depth 5 although other pickup stops are included afterwards.

Figure 4.2 illustrates the top of a search tree and shows the bit-vector implementations of the visited cities, the current time, premium service information, and the current capacity of the tour at each node. Note that some branches can be cut early if constraints are violated. For instance, the furthest to the right sub-tree can be ignored as the delivery of order 1 would be done before its pickup.

Table 4.1: Example for checking the capacity constraints when solving a problem with pickups and deliveries.

depth	χ_C	χ_M	is pickup stop	plan is valid (Eq. 4.10)
0	0	0	-	-
1	1	1	yes	true
2	2	2	yes	true
3	2	3	no	true
4	2	4	no	true
5	2	5	no	false
5	3	4	yes	true
6	4	4	yes	true

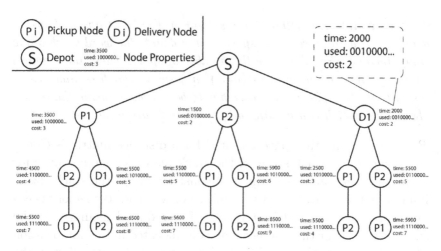

Figure 4.2: Example of the top of a search tree for solving constraint TSPs with pickups and deliveries.

Having pickups and deliveries simultaneously does not affect the optimality of the decision-making process (if the algorithm is not interrupted by exceeding a maximum number of expansions).

Theorem 4.3. *By setting the number of allowed expansions to infinity the solver is optimal for admissible lower bounds, the above pruning rules, and the objective functions specified in Eq. 4.7, Eq. 4.8, and Eq. 2.2.*

Proof. If no pruning were applied, every possible solution would be generated, so that the optimal solution would be found. Pruning rules that satisfy capacity and time-window constraints, cut off infeasible branches from the search tree so that the solution will be optimal. In addition, the search tree is only pruned by the upper bound u if the maximum depth is reached and all cities are still visited (to satisfy Eq. 4.7 and Eq. 4.8). If the tree is pruned by finding a better lower bound for the admissible objective functions, exploring the subtree cannot lead to better solutions than the one stored in u. □

Theorem 4.4. *Algorithm 1 for solving a TSP that has pickups or deliveries is done in constant space. Let w denote the word width of the computer and $n = |S|$ denote the number of stops. Then, the memory consumption is bounded by $O((n/w) \cdot n^2)$ space, which is allocated in the initialization phase. No space has to be allocated during the search. Heuristics, which can additionally be applied, are neglected.*

Proof. Assuming the space required for a distance matrix is $O(n^2)$. As shown above, each state on the stack requires $O(n/w)$ space for maintaining a constant number of constraints. As the stack size is limited by the depth times the number of successor, its memory needs are also bounded by $O((n/w) \cdot n^2)$. The required memory for the stack as well as for the distance matrix is allocated in the initialization phase. Consequently, no additional memory is allocated during the search.   □

Edelkamp and Gath (2013, pp. 252-253) showed that this DFBnB algorithm allows millions of expansions in less than a second, even if a *column-minimum heuristic* (Edelkamp and Gath, 2013, p. 250) is applied efficiently.[7] The solved TSPTWs with 20 customers are taken from the well-established benchmark set of Dumas et al. (1995). Depending on the constraints, most of the problems are solved optimally in less than a second or in a few seconds. The tests run on an Intel quad core i7 processor. As the memory consumption is constant, memory requirements can be neglected.

4.4.2 Time-Efficient Decision-Making

The DFBnB algorithm is not designed to solve large TSPs with more than approximately 20 stops.[8] In large problems, the calculation would not be finished in reasonable time or the calculation would

---

[7]Indeed, the effort to create the initial computation of the column-minima is done in $O(n^2)$. Updating the matrix by adding an additional stop can be performed in $O(n)$.

[8]The exact size of optimally solvable problems depends on the specific constraints and the effects of pruning.

have to be stopped once a maximum amount of expansions is reached. However, in the second case, it mostly fails to return a feasible solution. The reason is that the depth-first algorithm explores, e.g., the left branch first. In large problems the probability is high that hundreds of constraints are not fulfilled in this branch. This probability increases if pickup and delivery restrictions have to be considered. Indeed, the pruning rules cut the subtree if a constraint is violated. However, the algorithm continues to search for a solution within this unsatisfactory branch in a structured brute-force way. Although the pruning rules are applied as early as possible, it is very unlikely to find an adequate solution, because the number of unsatisfactory combinations remains unmanageable. In most cases, the maximum number of expansions is reached without finding a feasible solution. As a result, the computational complexity of the problem limits the algorithm to solve only small-size problems optimally.

To compute high-quality solutions in reasonable time for large problems as well, a novel approach was developed, which is based on Nested Monte-Carlo Search with Policy Adaptation also called Nested Rollout Policy Adaptation (NRPA). The algorithm is one of the first ones to apply the NRPA paradigm for solving large routing problems and especially problems involving pickups and deliveries, a domain which is generally dominated by Operations Research approaches (cf. Section 2.1).

Nested Monte-Carlo Search with Policy Adaptation

Monte-Carlo Search is a randomized search algorithm which iteratively performs random searches, so-called *rollouts*, within the search space, until the algorithm finds a valid solution, or a maximum amount of time has elapsed, or a maximum number of rollouts has been performed. The search method has particularly been applied to solve problems with a huge search space where no adequate lower and upper bounds are available. In contrast, *nested rollouts* perform an additional heuristic that determines next moves within the rollouts, to guide the search (Yan, Diaconis, Rusmevichientong, and

Roy, 2004). In further applications, this heuristic has successively been improved for solving challenging combinatorial problems such as Klondike Solitaire (Bjarnason, Fern, and Tadepalli, 2009). Nested Monte-Carlo Search (NMCS) (Cazenave, 2009) extends this approach by the concept of *levels*. At each level l, for each possible move (decision) at the current node of the decision tree the algorithm performs nested rollouts at level $l - 1$. Recursively, level $l - 1$ investigates all possible successor moves of the selected move in level l. If level $l = 0$, the search executes a random rollout. The best result at each level is propagated to the higher level, to identify and choose the best move found. Consequently, at each level, the best result has to be saved, because searches at lower levels may find worse solutions.

NMCS has been further extended by adapting a policy and introducing the concept of *iterations* (Rosin, 2011). The policy is used during the search to guide the search in promising directions. Thus, if n denotes the number of iterations, the algorithm performs n searches at each level. While the NRPA investigates all possible moves in depth $d = l$ of the decision tree at level $l - 1$, Nested Rollout Policy Adaptation (NRPA) executes n nested searches at level $l - 1$, which all start at the root of the decision tree and follow a policy until they reach a leaf. After a search at level $l - 1$ has been performed, the results are evaluated at level l and the policy is adapted by the best solutions found in level $l - 1$. When adapting the policy, a learning rate α adjusts the impact of the solutions found in the lower level. The algorithm is successfully applied for solving Crossword Puzzles and Morpion Solitaire (Rosin, 2011). A comprehensive survey about Monte-Carlo search in general is provided by Browne, Powley, Whitehouse, Lucas, Cowling, Rohlfshagen, Tavener, Perez, Samothrakis, and Colton (2012).

Recently, NRPA search has also been applied to efficiently solve the well-known Traveling Salesman Problem with time windows (TSPTW) optimally or very close to optimally for small problem instances with up to 50 cities (Cazenave and Teytaud, 2012). As described above, at level $l = 0$ of the recursive search a nested rollout is invoked. The likelihood of choosing a move within a rollout is determined by the policy and by three additional well-known heuristics for Vehicle Routing Problems, which are derived from Solomon (1987). Therefore, the policy considers

1. the distance from the last city to the next city,

2. the amount of wasted time, if a city is visited too early, and

3. the remaining time until the latest possible visiting time of a following city.

Moreover, Cazenave and Teytaud (2012) extended domain specific knowledge to solve TSPTWs with NRPA by restricting the possible successors (moves) within the rollout function. They force the algorithm to choose cities next that have reached the end of their latest possible visiting time. The general idea is to visit these cities next, because they have to be visited anyway and this has to be taken into account within further rollouts. In addition, the algorithm only considers cities that allow visiting other cities afterwards.

If hard constraints such as time windows and capacity constraints are not satisfied within a tour, the costs increase significantly for each violation. In this case, the result of a rollout is a constraint violating tour. Consequently, the algorithm minimizes constraint violations with highest priority. Details of the algorithm for solving TSPTWs are provided by Cazenave and Teytaud (2012). Algorithmic refinements to accelerate the search are given by Edelkamp et al. (2013).

NRPA for solving TSPs with Pickups and Deliveries

For solving large TSPs with pickups and deliveries and additional constraints such as time windows, handling times, and capacity constraints (cf. Section 2.1), the next algorithm extends the NRPA algorithm.

Firstly, the investigation focuses on the special case that the delivery stop has to be visited immediately after the pickup stop. Then, the problem is also called Stacker Crane Problem (SCP) (Srour and van de Velde, 2013). Consequently, a vehicle can only serve a single order at a time which allows neglecting capacity constraints. Moreover, it is possible to map the SCP to an asymmetric TSP as follows: the shortest-path length from each delivery location d_o of an order $o \in O$, is connected to all pickup locations p_o. As a result, a matrix of the size $(|O|+1) \cdot (|O|+1)$ is computed which can be used as input for any TSP solver such as the NRPA algorithm of Edelkamp et al. (2013).

In the more general TSP with pickups and deliveries, the constraint that the pickup stop of an order is immediately followed by the delivery stop is relaxed. The only constraint is that the pickup stop of an order has to be visited before the delivery stop (cf. Eq. 2.11). Consequently, the size of the distance matrix increases to $(2|O|+1) \cdot (2|O|+1)$ and the complexity of the problem grows significantly.

The recursive *search* function shown in Algorithm 2 is the main procedure of the NRPA algorithm for solving large TSPs with Pickups and Deliveries and additional constraints. The level-specific policy at each level l is adapted if a better solution at level $l-1$ has been found. As soon as all iterations have been executed, the global policy, which is applied by the *rollout* function, is overwritten. If the algorithm reaches level 0, the search function performs a rollout.

The *rollout* function is the most cost-intensive procedure of the NRPA algorithm. It samples a tour from the root of the search tree until a solution (a complete tour) has been found at a leaf by following the global policy. Algorithm 3 shows the implementation of the *rollout* function which extends the rollout function of the TSPTW solvers of Cazenave and Teytaud (2012) and Edelkamp, Gath, Cazenave, and Teytaud (2013). By setting flags, already visited successors are eliminated from the set of possible successors, so that any generated solution is a permutation of stops.

In contrast to the original implementation provided by Cazenave and Teytaud (2012), the TSP-related refinement to enforce a specific successor if a violation is certain is deactivated for the computation of the successor set. The reason for this is as follows. In small-size problems, choosing the first-fail strategy might accelerate the search because bad moves are identified early and are prevented from being repeated by the policy in further runs. However, in medium or large-

---

Algorithm 2 NRPA search function

1: **procedure** SEARCH(*level*,*iterations*)
2:     $best.score$ ← MAX_VALUE
3:     **if** *level* = 0 **then**
4:         $result$ ← ROLLOUT()
5:         $best.score$ ← $result.eval$
6:         $best.tour$ ← $result.tour$
7:     **else**
8:         $policy[level]$ ← $polGlobal$ // polGlobal denotes the global policy
9:         **for all** $i \in \{1, ..., iterations\}$ **do**
10:            r ← SEARCH($level - 1$, *iterations*)
11:            **if** $r.score < best.score$ **then**
12:                $best.score$ ← $r.score$
13:                $best.tour$ ← $r.tour$
14:                ADAPT($best.tour$, $level$)
15:            **end if**
16:         **end for**
17:         $polGlobal$ ← $policy[level]$
18:     **end if**
19: **return** $best$
20: **end procedure**

---

size problems it is preferable to improve the policy first, because in the exponentially growing state space the first-fail strategy of Cazenave and Teytaud (2012) would consume too much effort until all failures have been included in the policy and there is less time left to further improve the policy by good solutions. First test-runs in larger problems revealed a significant improvement of the solution quality as well as of the time performance without the first-fail strategy. Thus, the successor set is only determined by applying the following heuristic (lines 7-21). Firstly, the *check(i)* method checks for an unvisited successor i if it satisfies the hard constraints. In our case, it forces every order's pickup to precede its delivery. In addition, the procedure prevents a possible successor i, which had already failed being inserted in the tour, is investigated again. Therefore, it memorizes visited stops and also checked but unvisited stops. This is also necessary because the objective function has changed (cf. Eq. 4.7 and Eq. 4.8) in contrast to the standard TSP (cf. Definition 2.1). Thus, since it is not ensured that there is a tour which includes all stops without violating any constraints, it is possible that there remain unvisited stops. If a possible successor node violates the hard constraints, it is ignored and the number of violations is increased (line 30). Next, for each remaining successor the procedure examines if there is another unvisited successor j which cannot be visited after i. If the test is positive, this i is also removed from the set of possible successors.

In case the successor set remains empty, an arbitrary tour permutation is applied which considers only the precedences between pickup and delivery stops (lines 22-28). If the successor set is still empty because precedences are not satisfied, the number of violations is increased respectively (lines 29-32). Therefore, dispensable checks are avoided and the evaluation of this run is accelerated. When the set of possible successors is fixed, a random choice based on the current global policy is applied (lines 33-42). The biased choice of a successor refers to (roulette wheel) fitness selection in genetic algorithms. If the successor has been determined, the tour is extended by the respective stop (either pickup or delivery location), all further constraints are checked and violations are counted (lines 43-55). Next, the while loop continues in line 3 and processes the next stop. Finally, the number of violations is considered by the cost evaluation which is returned by the NRPA *search* procedure. Therefore, each violation is scaled with a constant (10^6 in this example).

As the most cost-intensive *rollout* procedure has to be called $iterations^{level}$ times in total, it is reasonable to stop the algorithm if no time is left or a maximum number of iterations is reached. In this case, the best global solution found has to be maintained, e.g., by a procedure *saveGloablBest(best)* in line 13-14 in Algorithm 2.

Further simplifications are applied but are not included in the pseudo-code in Algorithm 3 to keep the algorithm as general as possible. For instance, as in the case study presented in Chapter 7 and in the problems of the benchmarks presented in this chapter, it is not necessary to return to the depot, the makespan ms (the

---

Algorithm 3 NRPA rollout function for TSPs with Pickups and Deliveries

---

1: **procedure** ROLLOUT
2:    $makespan \leftarrow cost \leftarrow cap \leftarrow city \leftarrow 0$
3:    $tourSize \leftarrow 1$ // start the tour at the first stop(current position)
4:    $N \leftarrow |Stops| - 1$ // Stops is a global variable
5:    **while** $tourSize < N + 1$ **do**
6:       $sum \leftarrow s \leftarrow 0$
7:       **for all** $i \in \{1, ..., N\}$ **do**
8:          **if** $\neg visited[i] \wedge$ CHECK(i) **then** //visited is a global array
9:             $successors[s++] \leftarrow i$
10:             **for all** $j \in \{1, ..., N\}$ **do**
11:                **if** $\neg visited[j] \wedge i \neq j \wedge$ CHECK(j) **then**
12:                   //earliestTime/latestTime are global const. arrays
13:                   **if** $earliestTime[i] > latestTime[j] \vee$
14:                     $makespan + dist[city][i] > latestTime[j]$ **then**
15:                     $s--$
16:                     **break**
17:                  **end if**
18:                **end if**
19:             **end for**
20:          **end if**
21:       **end for**
22:       **if** $s = 0$ **then**
23:          **for all** $i \in \{1, ..., N\}$ **do**
24:             **if** $\neg visited[i] \wedge$ CHECK(i) **then**
25:                $successors[s++] \leftarrow i$
26:             **end if**
27:          **end for**
28:       **end if**
29:       **if** $s = 0$ **then**
30:          $violations \leftarrow violations + N - tourSize$
31:          **break**
32:       **end if**

33:     **for all** $i \in \{1, ..., N\}$ **do**
34:         $value[i] \leftarrow e^{(policy[city][successors[i]])}$
35:         $sum \leftarrow sum + value[i]$
36:     **end for**
37:     $random \leftarrow \text{RAND}(0, sum - 1)$
38:     $i \leftarrow 0$
39:     $sum \leftarrow value[0]$
40:     **while** $sum < random$ **do**
41:         $sum \leftarrow sum + value[++i]$
42:     **end while**
43:     $prev \leftarrow city$
44:     $city \leftarrow successors[i]$
45:     $tour[tourSize++] \leftarrow city$
46:     $visited[city] \leftarrow true$
47:     $cost \leftarrow cost + dist[prev][city]$
48:     $makespan \leftarrow \text{MAX}(makespan + dist[prev][city], earliestTime[city])$
49:     $cap \leftarrow weight[city]$
50:     **if** $cap > maxCap$ **then** //maxCap is a global constant
51:         viol++
52:     **end if**
53:     **if** $makespan > latestTime[city]$ **then**
54:         $violation++$
55:     **end if**
56:     **end while**
57: **return** $(10^6 \cdot violation + cost, tour)$
58: **end procedure**

accumulated time at which the stop is visited) can be optimized instead of the costs (because the objective function is in general to minimize the time and not the driven distance).

Especially in the real-world problems investigated in this thesis, it is impossible to find a feasible solution which includes all orders without any violation. As described above, the objective function of these problems changes. The goal is to maximize the number of transported orders (minimize the violations) and to minimize the total distance. In this case, lines 43-55 have to be changed. The algorithm only extends the tour if the constraints of the new stops are satisfied. Otherwise, neither the makespan nor the capacity of the vehicle is increased and also the respective delivery stop is remarked as visited. If a delivery stop cannot be included, the respective pickup stop s of the tour $(s_0, ..., s_l)$ with $s_i \neq s_j$ at position $k < l$ has to be removed from the current tour. Consequently, the makespan has to be adapted. To this end, the distance between t_{k-1} and t_k as well as the distance between t_k and t_{k+1} is substituted by the direct shortcut between t_{k-1} and t_{k+1} (as long as the triangle equation between cities is fulfilled - which is assumed - this never increases the makespan). The reduced capacity is considered in the ongoing computation.

Algorithm 4 depicts the *adapt* procedure which is similar to the policy adaptation of Rosin (2011, p. 650). It updates the level policy if the *search* function returns a better solutions than the current best one.[9] While the policy of Rosin (2011, p. 650) evaluates a chosen action in a specific state, the policy in Algorithm 4 assesses the "goodness" to visit a specific stop after the current stop. More in detail, following the current global policy $\pi : (S \times S) \setminus \{(s, s) \in S \times S\} \to \mathbb{R}$, the rollout chooses after the current stop $s \in S$ the successor stop $s' \in S$ wrt. $e^{\pi(s,s')}$ (thus, the policy π is a matrix $M : \{1, ..., |S|\} \times \{1, ..., |S|\} \to a_{i,j}$ which defines for each $(i, j) \mapsto a_{i,j}$ with $i \neq j$). The exponential function e is taken to reduce the impact of the policy and to avoid overfitting. As the entire stop-to-stop table would exceed the memory requirements, it is limited to the essential part to be learned, which is the measurement for going from one stop to the next. Therefore, the *policy adaption* performs gradient descent as follows.

---

[9]Note that due to increase the exploitation of the nested search, the best solution in the recursive procedure does not refer to the globally best solution.

Algorithm 4 NRPA adapt function

1: **procedure** ADAPT(*tour*, *level*)
2:     $N \leftarrow |Stops| - 1$ // Stops is the global set of stops
3:     **for all** $i \in \{1, ..., N\}$ **do**
4:         $checked[i] \leftarrow false$
5:     **end for**
6:     $node \leftarrow 0$ // assuming the vehicle starts at the first stop
7:     **for all** $p \in \{1, ..., N\}$ **do**
8:         $successors \leftarrow 0$
9:         **for all** $i \in \{1, ..., N\}$ **do**
10:            **if** $\neg checked[i]$ **then**
11:                $moves[successors++] \leftarrow i$
12:            **end if**
13:        **end for**
14:        $policy[level][node][tour[p]] \leftarrow policy[level][node][tour[p]] + \alpha$
15:        $z \leftarrow 0.0$
16:        **for all** $i \in \{1, ..., successors\}$ **do**
17:            $z \leftarrow z + e^{(polGlobal[node][moves[i]])}$
18:        **end for**
19:        **for all** $i \in \{1, ..., successors\}$ **do**
20:            $policy[level][node][moves[i]] \leftarrow policy[level][node][moves[i]] -$
            $e^{polGlobal[node][moves[i]])}/z \cdot \alpha$
21:        **end for**
22:        $node \leftarrow tour[p]$
23:        $checked[node] \leftarrow true$
24:    **end for**
25: **end procedure**

The sequence of successor cities $s' = (s'_0, \ldots, s'_l)$ of $s = (s_0, \ldots, s_l)$ with $s_{i+1} = s'_i$ has probability

$$P(s, s') = \frac{\prod_{j=0}^{l} e^{\pi(s_j, s'_j)}}{\sum_{i=0}^{l} e^{\pi(s_j, s'_i)}}. \tag{4.11}$$

The gradient of the logarithm at j is $1 - e^{\pi(s_j, s'_j)}/\sum_{i=0}^{l} e^{\pi(s_j, s_i)}$, so that we add α to the selected successor city (line 14) and subtract $e^{\pi(s_j, s'_j)}/\sum_{i=0}^{l} e^{\pi(s_j, s'_i)}$ from the others (lines 16-21). This ensures that policy adaptation increases the probability of the established tour, which outperformed the current best tour.

To further accelerate the search, the global policy π is initialized with $\pi_{init} : (S \times S) \setminus \{(s, s) \in S \times S\} \to [0, 1] \subset \mathbb{R}$ wrt. the shortest-path distances between stops as follows. As above let $S = \{s_1, \ldots, s_n\}$ with $n > 1$ denote all stops. For all $(s, s') \in (S \times S)$ with $s \neq s'$ the algorithm computes the value of $\pi_{init}(s, s')$. For this purpose, let $l \in ((S \times S)^{n-1})^n$ denote a list of sorted lists defined as

$$l = (l_1, \ldots, l_n) \tag{4.12}$$

with

$$l_i = ((s_i, t_{i,1}), \ldots, (s_i, t_{i,n-1}))1 \leq i \leq n \tag{4.13}$$

$$\forall_{1 \leq j \leq n-1} : t_{i,j} \neq s_i \tag{4.14}$$

with the ordering

$$\forall_{1 \leq i < n-1} : dist(s_i, t_{i,j}) \leq dist(s_i, t_{i,j+1}) \tag{4.15}$$

and with

$$\{t_{i,1}, \ldots, t_{i,n-1}\} = S \setminus \{s_i\}. \tag{4.16}$$

[10] Moreover, let v denote the finite series $v = v_1, \ldots, v_{n-1}$ with equidistant values

$$v_k = \begin{cases} 1 & \text{if } n = 2 \\ \frac{i-1}{n-2} & \text{otherwise} \end{cases} 1 \leq k \leq n - 1. \tag{4.17}$$

Then, $\pi_{init,i} : (S \times S) \setminus \{(s, s) \in S \times S\} \to [0, 1] \subset \mathbb{R}$ maps each stop pair to equidistant weights between 0 and 1 by setting $(s_i, t_{i,j})$ to v_j with $1 \leq i \leq n$ and $1 \leq j \leq n - 1$.

Tests have revealed that it is suitable to limit the initial values to the interval between 0 and 1 to avoid overfitting the policy. As distances between cities are not unique, the ordering by Eq. 4.15 is not deterministic. However, this is neglected because NRPA is an inherently random algorithm (cf. line 37 in Algorithm 3).

---

[10]Note that $t_j = s_j$ is not necessarily true.

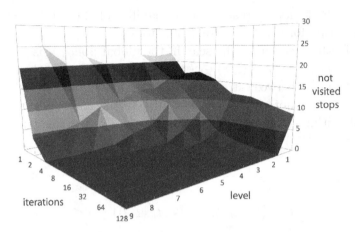

Figure 4.3: The solution quality for the problem with 20 orders measured by the average number of violated constraints of 10
runs for each *level* and *iteration* pair.

Parameter Configuration

In general, two parameters affect the solution quality in NRPA: the
number of *iterations* and the *level*. This section investigates the
effects of varying these parameters to identify suitable configurations
for application.

In order to derive pertinent conclusions from the results, each
experiment is repeated 10 times and average values of all runs are
recorded. Due to the fact that rollouts are the most cost-expensive
operations within the algorithm and the complete run requires performing *iterations*[level] rollouts in total, the algorithm terminates at the
latest after performing 40,000 rollouts. Note that the algorithm has
no knowledge about the best or optimal solution. Even if it finds the
optimal solution quite early, it continues the search until all rollouts
are performed or a termination is forced.

The first investigation focuses on a problem with 20 orders (thus,
40 stops have to be visited). The problem instance is taken from the

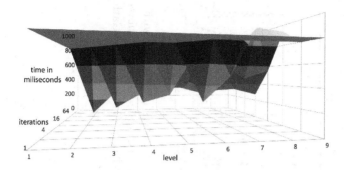

Figure 4.4: The average computation time of 10 runs for each *level* and *iteration* pair for the results of the problem with 20 orders (which is shown in Figure 4.3).

benchmark of Jih and Hsu (2004).[11] Figure 4.3 shows the solution quality measured by the average number of violated constraints. The results show that a minimum amount of rollouts have to be performed to compute suitable results. Thus, feasible solutions are guaranteed with 4 iterations and at least 8 levels (max. $4^8 = 65,536$ rollouts) or at least 64 iterations and 2 levels (max. $64^2 = 4,096$ rollouts). Consequently, Figure 4.4 investigates the computation time for the first-found best solution, which refers to the respective solution in Figure 4.3. Figure 4.4 reveals that a level 2 search with 64 iterations determines feasible solutions in less than 100 milliseconds on average while increasing the level leads to higher computation times in this small-size problem. In addition, the results prove that there is no significant difference in the solution quality if the number of levels is increased and the search is terminated after a fixed number of rollouts. This is obvious because, e.g., a level 4 search is completely included in the level 5 search. As a result, if the algorithm already terminates in the level 4 search, the level 5 search is never executed. In this case, further increasing the level has no impact.

---

[11]The benchmark is available at http://wrjih.wordpress.com/2006/12/09/pdptw-test-data/ (cited: 1.9.15).

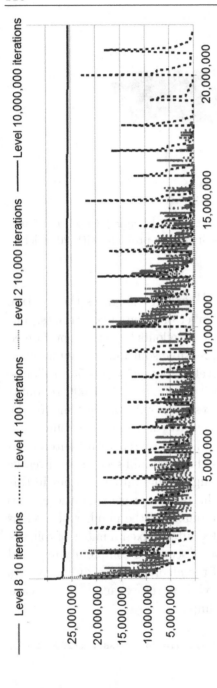

Figure 4.5: The learning curves for solution quality of the benchmark with 100 orders (200 stops) with different NRPA parameters (the x-axis shows the number of rollouts, while the y-axis shows the change in solution quality for each improvement at a certain level) (Edelkamp and Gath, 2014, p. 28).

Another experiment by Edelkamp and Gath (2014) examines several configurations to solve the benchmark instance of Jih and Hsu (2004) which includes 100 orders. Figure 4.5 depicts a cross-comparison of different parameters. The figure plots the number of runs (x-axis) and the respective solution quality (y-axis). The best results for the benchmark instance with 100 orders are obtained with level 2 and 10,000 iterations, while no solution is computed with level 16 and 3 iterations (even after hours of waiting) and level 1 with 100,000,000 iterations.

The next experiment focuses on a large problem. Figure 4.6 compares a varying number of iterations of level 2 and level 3 searches for the benchmark with 200 orders (400 stops) provided by Jih and Hsu (2004). Figure 4.6a reveals that at least 600 iterations are required to determine feasible solutions including all orders with a high probability. In contrast to small-size problems, this is caused by the increased complexity. While the number of iterations determines the exploitation rate, the level affects the exploration of the search space. In difficult problems, more exploitation is required to determine more qualitative solutions. Only qualitative solutions lead to a qualitative adaption of the global policy. If not enough iterations are applied, the adaptation of the global policy is not sufficiently goal directed.

Figure 4.6b describes the score of feasible solutions and Figure 4.6c depicts the computation time of the first-found best solution. All runs are terminated after three hours at the latest. As the number of iterations is low, it is obvious that the level 3 search outperforms the level 2 search, because level 2 searches terminate earlier while level 3 searches continue looking for better solutions. If the number of iterations is sufficiently large to determine qualitative solutions, Figure 4.6c also indicates that level 2 is slightly preferred to level 3. As more qualitative solutions are determined, a higher level allows more exploration than a lower level. This prevents getting stuck in local minima.

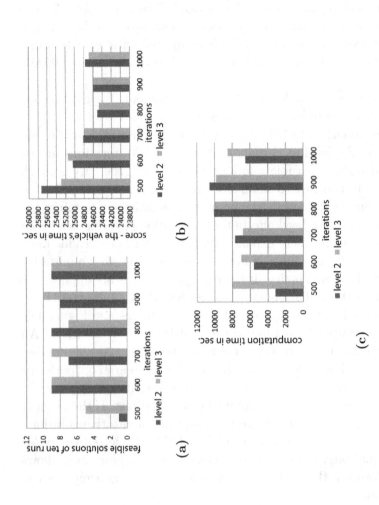

Figure 4.6: Feasible solutions, the respective average solution quality, and the average time for computing the first best solutions for level 2 and level 3 searches with a varying number of iterations.

Table 4.2: The results of the NRPA algorithm with 3 levels and 128 iterations performed on Jhi's PDP benchmark (Jih and Hsu, 2004).

Instance	Value		Percentage		CPU Time	
	Best	Median	Valid	Best	Average	Worst-Case
PDP 20	2,031	2,031	10/10	10/10	0.476s	3.028s
PDP 60	5,658	5,658	7/10	7/10	100.436s	349.533s
PDP 80	7,849	7,849	10/10	10/10	101.654s	273.658s
PDP 100	10,600	10,600	10/10	10/10	495.767s	790.347s

After the investigation of a small (20 orders) and two large-size problems (100 and 200 orders), further problem instances with 60, 80, and 100 orders reveal similar outcomes that strengthen the presented results. In conclusion, decreasing the search level and increasing the number of iterations leads to better results for solving difficult TSP with pickups and deliveries and additional constraints with NRPA within short computational time. Indeed, at both ends of the spectrum, i.e., NRPA with level 1 and level 16 do not find any feasible solution (cf. Figure 4.5). The reason is that an NRPA with level 1 is a series of pure rollouts (comparable to a greedy search), and policy learning is too weak to direct the search towards finding feasible solutions, while an NRPA with level 16 simply forgets too much about previous trials. This case demonstrates the classic extremes of exploration and exploitation. In the general, however, to determine qualitative solutions it is more relevant to adapt the global policy qualitatively instead of guiding the search towards new directions and adapting the global policy by lower quality solutions. The results substantiate this for the TSP with pickups and deliveries and with time-window constraints.

Benchmark Evaluation on Small and Medium-size Problems

The following experiments ran on an Intel(R) Core(TM) i7-2620M CPU at 2.7 GHz. The computer is equipped with 16 GB RAM. The

Table 4.3: The best results achieved with the MX2(FCGA) algorithm of
Jih and Hsu (2004) on thier PDP benchmark.

Instance	Value Best	Percentage Valid (180 runs)	Best	CPU Time ≈ Average (180 runs)
PDP 20	2,030	100	11/30	≤ 100s
PDP 60	5,658	87.22	30/30	300s
PDP 80	7,849	80.56	1/30	1,400s
PDP 100	10,600	63.89	26/30	2,300s

memory requirements of our NPRA implementation are insignificant
and dominated by the size of the policy. It is required to save a policy
for each level of the search. The algorithm is implemented in Java.

Table 4.2 shows the results of the NRPA algorithm evaluated with
the benchmark provided by Jih and Hsu (2004). Table 4.3 examines
the results achieved with the genetic algorithm (MX2(FCGA)) of Jih
and Hsu (2004) on the same benchmark.[12] The results show that
the solution quality of both algorithms is nearly identical (only for
the 20 PDP just a single score point is not achieved with NRPA).
However, applying NRPA is more reliable. In 37 out of 40 runs the
NRPA computed feasible solutions, while the genetic algorithm has
an average success rate of 82.92% for finding feasible solutions (after
performing 180 runs in total). Moreover, if the NRPA determines a
feasible solution, this this also the best-known or optimal solution. In
addition, the computation time of NRPA is significantly lower (but
note that the results of the genetic algorithm were computed several
years before and no hardware configuration is presented).

Next, the performance is compared to the algorithm provided
by Hosny and Mumford (2010). While the general quality of these
results is nearly the same, in some instances the presented solutions
are better than the optimal solutions determined with dynamic programing (Jih and Hsu, 2004). Thus, these results have to be wrong

---

[12] Jih and Hsu (2004) developed multiple generic algorithms for solving the
benchmark problems. The MX2(FCGA) algorithm is the one with the best results.

or at least inconsistent with the description in the text. In order to reproduce the error, the cost function in the NRPA approach is varied (e.g., by dropping waiting times at stops). However, it failed to identify the error of Hosny and Mumford (2010) by changing the assumptions.

In conclusion, NRPA determines state-of-the-art solutions for the TSP with pickups and deliveries and with time-window constraints. It has a higher success rate for finding feasible solutions. Thus, applying NRPA on this benchmark is more reliable than the established approaches presented by Jih and Hsu (2004) and Hosny and Mumford (2010).

Benchmark Evaluation on Large-size Problems

All experiments ran on an Intel(R) Core(TM) i5-2520M CPU at 2.5 GHz. The memory requirements are negligible (see above). For a sequence of 10 runs, Table 4.4 provides information on the best and the median solution costs. The median cost is recorded, because in case of constraint violation the average value is increasing rapidly. In addition, it shows the fraction of valid and best solutions found and provides the average number of runs required to obtain this best solution. The maximum number of runs is implicitly given by the limit of the search tree: for the level 2 NRPA with $2|S|$ iterations, the experiment terminates after $(2|S|)^2$ runs. The depicted runtimes are worst and average cases and enforced by either solving the problem with the best possible solution or by hitting the limited number of rollouts. First feasible solutions are usually established much earlier.

Table 4.4 proves that the algorithm solves problems with at least 200 orders (400 stops). If the results are compared with the solutions computed by the most efficient approach (a genetic algorithm) provided by Hosny and Mumford (2007, p. 21), the NRPA computes solutions with a similar quality. Also the success rate for finding optimal and feasible solutions is approximately equal. However, also in this benchmark set in some instances Hosney's genetic algorithm finds solutions which are better than the optimal solutions computed

Table 4.4: Results of 10 runs applying the NRPA algorithm with level 2 and 2N iterations in Hosny's PDP Benchmark (Edelkamp and Gath, 2014, p. 28).

Stops $|S|$	Value		Percentage		Rollouts for Best		CPU Time	
	Best	Median	Valid	Best Solution	Average	Maximum	Worst-Case	Average
201	9,757.00	9,757.00	8/10	8/10	102,067	161,604	6m3s	3m36s
221	11,641.03	11,641.03	10/10	10/10	144622	195,364	8m32s	6m5s
241	12,143.00	12,143.00	10/10	10/10	96,217	232,324	7m4s	5m21s
261	14,057.00	14,057.00	10/10	10/10	146,995	272,484	17m39s	9m6s
281	15,111.00	15,160.12	8/10	5/10	271,614	315,844	25m14s	20m38s
301	16,976.00	16,976.00	10/10	7/10	190,111	362,404	23m48s	16m36s
321	18,167.00	18,167.00	8/10	7/10	316,842	412,164	45m35s	34m33s
341	19,924.00	19,924.00	10/10	10/10	246,747	465,124	55m11s	28m46s
361	22,107.29	22,107.29	10/10	10/10	244,314	521,284	42m37s	30m16s
381	23,826.00	23,826.00	10/10	10/10	270,967	580,644	61m55s	35m25s
401	24,184.00	24,198.86	8/10	4/10	587,384	643,204	128m	111m

with dynamic programming approaches by Jih and Hsu (2004) (see above). Thus, small deviations from the solutions of the generic algorithm may be caused by this.

Ultimately, the results prove that NRPA computes valid and state-of-the-art results for large TSPs with pickups, deliveries, and time windows. The quality of the results as well as the success rate for finding feasible and best-known solutions reveal that NRPA can compete with to other heuristic and meta-heuristic approaches.

Iterative Widening

Further experiments by Edelkamp and Gath (2014) investigate the performance of a parameter-free implementation of NPRA which applies *iterative widening*. Instead of imposing a fixed number of iterations for each instance, the number of iterations is gradually increased within the search process. The obvious choice is to use odd numbers yielding the sequence of squares 1, 4, 9, 16, ..., k^2 for the number of iterations. The idea is to quickly find first solutions, quickly propagate them bottom-up, and put more effort into finding better solutions later. Other functions such as Luby sequences (Luby, Sinclair, and Zuckerman, 1993) (successful in SAT (Een, Mishchenko, and Srensson, 2007) solving) are available, but have not yet been tested.

The experimental results for this approach are indeed promising. For instance, in another 10-fold repeated experiment of Hosny's 100 PDP problem (cf. Table 4.4 first row) the optimal solution of 9,757 was found in all cases and obtained even faster within 3m13s (vs. 6m3s) in the worst case, and within 2m58s (vs. 3m36s) in the average case.

In feasible solutions (8 of 10 computed tours), the average number of applied rollouts is decreased from 45,828 to only 16,574 by applying the parameter-free implementation (in the worst case the number of required rollouts is reduced from 108,138 to 23,278). Thus, considering this problem of Table 4.4, the parameter-free method is three times faster than the NRPA configured with level 2 and 401 iterations.

Conclusion and Discussion

This section presented a novel approach for solving TSPs with pickups and deliveries as well as with time-window and capacity constraints by applying Nested Monte-Carlo Search with Policy Adaptation. In order to evaluate the approach and to determine suitable parameter values for the *level* and the number of *iterations*, first experiments investigate numerous instances with varying configurations. small-size instances are retrieved from the benchmark set provided by Jih and Hsu (2004), while the medium-size instances are taken from Hosny and Mumford (2007).

While the number of iterations determines the exploitation rate, the level affects the exploration of the search space. In difficult problems, more exploitation is required to achieve qualitative solutions. Only qualitative solutions lead to a qualitative adaption of the global policy. If the number of iterations is too small, the adaptation of the global policy is not sufficiently goal directed. The results show that decreasing the search level and increasing the number of iterations leads to better results for solving complex TSPs with pickups, deliveries, and additional constraints with NRPA within short time. Thus, it is more relevant to determine qualitative solutions to adapt the global policy qualitatively instead of guiding the search towards new directions and adapting the global policy by lower quality solutions.

From a machine learning point of view the object to be learned is a policy (that generalizes from the state-to-state estimates) which reflects the linkage of stops, i.e., which one is the best to visit next in a short tour. This gradually improved knowledge starting with some measure of initial distances is highly relevant in guiding the search process to improve tours: knowing that a particular stop is a good successor of another one is essential not only for a single but for many other tours as well. Thus, less adaptations to the level policy are needed to drive the solver towards better solutions.

Finally, the results show that NRPA is competitive with other heuristics and meta-heuristics such as genetic algorithms, at least in problems with 20 and 200 orders (up to 400 stops). In general, this problem size is sufficient for most real-world problems and especially for the decision-making process of the agents described above. NRPA computes state-of-the-art solutions and has a high rate of success for finding feasible and best-known solutions. With a limited amount of domain-specific information (only a single heuristic is applied in the rollout function) the algorithm handles problems in adequate computation time.

Despite of extensive analysis to determine pertinent parameter configurations for TSPs with pickups and deliveries, initial experiments show that applying a 2 level search and an iterative widening strategy is indeed promising.

Moreover, especially the dispAgent approach profits from the NRPA algorithm because the MAS allows to save and memorize policies. The vehicle agent applies the NRPA algorithm to compute the cost for an additional order. After each computation, it saves the learned policy. In the next computation, most stops of the tours are identical. Only the new stops of additional orders are appended. Thus, the vehicle agent initializes the policy with the old one and continues to improve it. Thus, the agent preserves the existing knowledge.

Another key advantage of the NRPA algorithm is that only the structure of the policy π and *rollout* procedure (cf. Algorithm 3) is domain-dependent. To this end, the algorithm can also be applied to solve other logistics problems such as the packing problem (Edelkamp et al., 2014). For instance, in domains where goods must be transported over large distances, the optimization of the vehicle's volume is highly relevant. In this case, the vehicle agent could explicitly use the NRPA algorithm to solve the packing problem and consider the result in its decision-making processes.

4.5 Parallel Shortest-Path Searches

In the established and often applied benchmarks for VRPs and TSPs, the distances between locations are Euclidian distances. These can be determined with low computational requirements. However, shortest-path searches on real-world infrastructure networks are cost-intensive operations. Therefore, efficient shortest-path searches are essential for multiagent-based transport processes. In particular for the decision-making of the vehicle agents, $|S| \times |S|$ shortest-path searches are required to compute the distances matrix of all-pairs shortest paths for $|S|$ stops. Despite the considerable importance of shortest-path searches, none of the existing multiagent-based approaches presented in Section 3.4 focuses on this problem.

This section compares the application of established and high-speed algorithms. It investigates the effects of shortest-path computations in multiagent-based negotiations. Next, the section focuses on the parallel application of a state-of-the-art *hub labeling* algorithm which is combined with *contraction hierarchies*. Several modeling approaches are compared and discussed. The goal is to increase the performance of the decision-making process with regard to runtime and memory consumption by the efficient implementation of shortest-path searches and the use of a suitable agent model. This is crucial for the implementation of autonomous control in real-world transport processes as well as for large scale logistics multiagent simulations.

4.5.1 The Impact of Shortest-Path Searches

Three well-established shortest-path algorithms are implemented for this investigation: the classic Dijkstra algorithm (Dijkstra, 1959) and an implementation of the A* algorithm, both of which use *radix-heaps* (Greulich et al., 2013) for an efficient graph representation, as well as a shortest-path algorithm which combines *hub labeling* with *contraction hierarchies*.

Since 2011, hub labeling algorithms (Abraham, Delling, Goldberg, and Werneck, 2011) in combination with contraction hierarchies (CH) (Geisberger, Sanders, Schultes, and Delling, 2008) have been known to be almost the most efficient shortest-path algorithms. For instance, shortest-path queries on the whole transport network of Western Europe are processed in less than a millisecond (Abraham, Delling, Goldberg, and Werneck, 2012, p. 34).

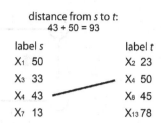

Figure 4.7: This example shows how the distance between a start node s and a target node t is determined by comparing their labels.

The idea of distance labeling algorithms is that the distance between two nodes is only determined by of comparing their assigned labels, which are ideally computed offline (cf. Figure 4.7). Therefore, search queries on the pre-computed labels can efficiently be performed online. In our implementation, the hub labels contain a list of references to multiple other nodes (the hubs). Within the construction process of the labels, the so-called *cover property* has to be satisfied, which means that both labels of any two vertices s and t must contain the same vertex that is on the shortest $s - t$ path (Abraham et al., 2012, pp. 25-26). The cover property guarantees that all shortest paths in a graph can be determined by the labels of the source and target nodes. The challenge is to create memory-efficient labels that satisfy the cover property. Applying the labeling algorithm on nodes which are saved in a CH, allows for memory-efficient label representations. In order to build the CH, the original graph g is extended to a larger graph g' which contains direct shortcuts between nodes instead of shortest paths in g. The algorithm iterates over all nodes and saves each node in the next higher level of the hierarchy. In this process, it calculates possible shortest-path shortcuts to other nodes. Therefore, the current node is considered to be removed from the graph and it is checked if all other shortest paths would still be included in the graph without this node. If a shortest path originally passed the *removed* node, a new shortcut is created to retain this shortest path. The general steps are shown in Figure 4.8.

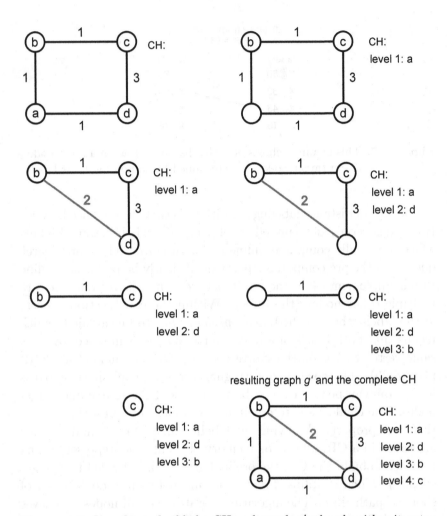

resulting graph g' and the complete CH

Figure 4.8: In order to build the CH and graph g', the algorithm iterates over all nodes and saves each node at the next higher level of the hierarchy. Therefore, it removes the respective node from the graph. If a shortest path originally passed the *removed* node, a new shortcut is created in g' to retain this shortest path.

The performance of the algorithm highly depends on the sequence of nodes in which they are added to the CH (Geisberger et al., 2008). An optimal sequence minimizes the search space for the optimal number of shortcuts. As this problem is NP-hard (Bauer, Columbus, Katz, Krug, and Wagner, 2010), Geisberger et al. (2008, pp. 322-324) suggest to use the following approach. To determine the next node that will be processed, all unprocessed nodes are sorted by a priority value. The node with the highest priority is processed next. The priority value of a node mainly depends on the edge difference between the current graph and the graph containing the shortcuts that result from processing that node. After adding a new node to the CH, some priorities have to be updated, because in every iteration the graph might be extended by a new shortest path. Due to the fact that the computation of the priorities is a cost-intensive operation, the value is estimated. The better the sequence of the iterated/selected nodes, the less shortcuts are determined, and the more efficient is the memory consumption and the search on the CH. In addition, there are also approaches which can be applied to time-dependent graphs (Batz, Geisberger, Neubauer, and Sanders, 2010) or to dynamically changing graphs (Geisberger, Sanders, Schultes, and Vetter, 2012).

Next, the hub labels are computed on the basis of the CH and the extended graph g'. This process starts at the highest level of the CH. For each level (node) a label is created. The label contains all references and information about the shortest distance to nodes saved at higher levels. Figure 4.9 gives an example for the creation of hub labels on the basis of a CH and a graph g'. Further optimization techniques to reduce the memory are not implemented yet, but provided by Abraham et al. (2012).

The goal is to investigate the impact of shortest-path searches on the runtime in multiagent-based logistics transport processes by using the dispAgent software system. Therefore, several simulations of exactly the same scenario with the three different shortest-path algorithms were performed within PlaSMA (cf. Section 6.4.1). The simulated scenario is similar to the one which is presented in Section 6.4. The underlying real-world transport infrastructure, which is imported

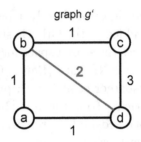

graph g'

CH:	hub labels:
level 1: a	
level 2: d	
level 3: b	
level 4: c	c: c,0

CH:	hub labels:
level 1: a	
level 2: d	
level 3: b	b: b,0; c;1
level 4: c	c: c,0

CH:	hub labels:
level 1: a	
level 2: d	d: d,0; b,2; c,3
level 3: b	b: b,0; c;1
level 4: c	c: c,0

CH:	hub labels:
level 1: a	a: a,0; b,1; d,1; c,2
level 2: d	d: d,0; b,2; c,3
level 3: b	b: b,0; c;1
level 4: c	c: c,0

Figure 4.9: The figure shows an example of how the hub labels are created on the basis of a CH and a graph g'. Starting at the highest level, for each level (node) a label is created. The label contains all references and information about the shortest distance to nodes saved at higher levels.

from OpenStreetMap[13], contains $85,633$ nodes and $196,647$ edges. All experiments were performed on a laptop computer with an Intel quad-core i72620-M CPU/2.7 GHz and 16 GB RAM.

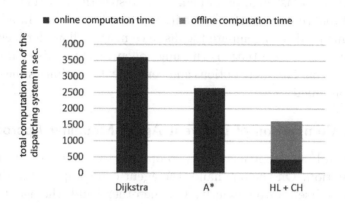

Figure 4.10: A comparison of the runtime of multiagent-based transport simulations. The only difference between the scenarios is the applied shortest-path algorithm.

Figure 4.10 compares the physical time which is required for the simulation of a scenario in PlaSMA. It clearly shows that the shortest-path algorithm significantly affects the runtime of the multiagent-based dispatching approach even on the relatively small infrastructures investigated in this experiment. Although the Dijkstra and A* algorithm are well-established and widely applied algorithms, Figure 4.10 shows that the shortest-path problem, whose complexity is in P, has a more significant impact on the system's runtime than solving the NP-hard TSP with a high-performance algorithm. Thus, if the well-established Dijkstra or A* algorithm is applied, it is clearly the most time-intensive operation of the multiagent-based dispatching system. This is not only true for the implemented MAS, but at least for most of the other MAS presented in Section 3.4 and Table 3.3, which are applied on real-world infrastructures, since the order agents have to compute a distance matrix first before they can solve the VRP-like sub-problems.

---

[13]http://www.openstreetmap.org (cited: 1.9.15).

In conclusion, it is essential to apply a high-performance shortest-path algorithm in multiagent-based applications for transport logistics such as hub labeling with contraction hierarchies. Only multiagent systems with fast high-performance shortest-path algorithms have an acceptable runtime performance in real-world scenarios. In addition, less time needed to compute the distance matrix allows for increasing the number of negotiations running concurrently. This enables the agents to validate more options and consequently optimizes the overall solution quality.

4.5.2 Comparison of Different Agent Modeling Approaches

In general, an agent has only two options to acquire shortest-path information. On the one hand, the agent can compute the shortest paths by itself autonomously. On the other hand, the agent might ask a *service provider* agent (a so-called routing agent) which then receives a routing request, computes the shortest path, and sends the result back to the agent. The goal of this investigation is to determine a suitable way of modeling for a scenario in which numerous routing requests have to be answered immediately (e.g., to compute distance matrices of cities or stops). Especially if the agents apply the hub labeling algorithm in combination with CH, it is not sufficient to consider the performance in handling search queries online, but also necessary to include the time for the creation of the hub labels and of the CH as well as the amount of memory which is required to save all the labels. The investigation of different agent modeling approaches was performed by Gath et al. (2015b) who analyzed the following cases.

In the first case, each agent has its full autonomy and maintains its own algorithm. Thus, each agent has to build and save the hub labels as well. This is time- and memory-intensive. In the second case, the agents outsource the memory and time consuming shortest-path operations to one or several routing agents. As a result, only the routing agent must build and save the hub labels. In this case, the optimal number of parallel running routing agents must be determined.

Beside the above mentioned options, Java programing language allows another approach that technically restricts an agent's autonomy slightly. It is possible to have the hub labels built by a single agent and save them in a static variable. While classic routing algorithms such as the Dijkstra algorithm manipulate the graph by saving distance information at the nodes to compute the shortest paths, the hub labeling algorithm performs read-only operations on the labels instead. Thus, the agents can directly access this static variable and perform the routing requests by themselves in parallel. Depending on the computer architecture the multiagent system is running on, these operations are performed also physically concurrently. However, this slightly violates the agent's autonomy, because all agents (running on the same Java Virtual Machine (JVM)) share the same component. Thus, they are technically not fully independent of each other.

The following investigation focuses on scenarios simulated in PlaSMA. In these scenarios 1,000,000 routing requests of several agents must be answered immediately (e.g., to compute distance matrices of stops/cities). To satisfy real-word requirements, the underlying transport infrastructure is the road network of Lichtenstein with 3,607 nodes and 8,401 edges. For the evaluation in reasonable time on conventional hardware, this area with a restricted number of nodes and edges has been chosen because it allows to pinpoint pertinent results by measuring average values of 10 runs in each setting. Nevertheless, the algorithm has successfully been applied to larger infrastructures with more than 300,000 edges and 200,000 nodes. The 1,000,000 search queries are requested by 50 agents. Thus, each agent asks for 20,000 shortest paths. The simulation is started with an agent, which generates 20,000 search queries with randomized start and end nodes for each of the 50 so-called consumer agents which are created. In order to guarantee the reproducibility of runs and an accurate comparison, the random seed of the random number generator is fixed in each experiment.

In Scenario 1, each consumer agent applies its own shortest-path algorithm. The agents start a pre-processing step to build up the hub labels as well as the CH and process all their queries by themselves.

In Scenario 2, the shortest-path algorithm is implemented by the *Singleton Pattern* (Gamma, Johnson, Helm, and Vlissides, 1994, pp. 127). In this scenario, there exists only a single instance of the algorithm on each JVM. All consumer agents process their queries by themselves, but operate on the same instance of the implemented algorithm saved in a static variable.[14]

In Scenario 3–12, between 1 and 50 routing agents are created. They maintain their own shortest-path algorithm and receive all the queries from the consumer agents by a FIPA-compliant ACL-message (cf. Section 3.2). Next, the routing agents compute the shortest path of the assigned requests and send an ACL-message with the answer back to the agent. The assignment of search requests to routing agents is uniformly distributed. The reproducibility and accurate comparison of this assignment is also ensured by applying a fixed random seed. All the simulations run on a notebook computer with an Intel quad-core i5-2500k processor, Windows 7 64 bit, and 16 GB RAM.

In each scenario, four performance indicators were measured. The most significant one is the total (physical) simulation time. Moreover, two performance indicators determine the time required for pre-processing. This is the earliest time an agent ends its pre-processing step and the elapsed time for all the agents to finish building the hub labels and the CH. The results are shown in Figure 4.11. The measured performance indicators are average values of 10 runs. As the shortest-path algorithm only performs read-only operations, the memory requirements of the whole MAS increase proportionally to the number of graph instances, which are created and saved concurrently by all the agents. This is obvious, but also rechecked in this investigation.

The results show that providing the *full* technical autonomy of agents requires higher memory utilization and longer runtime. Each agent must create and maintain its own routing algorithm. On a quad-core architecture, this process cannot be parallelized physically to 50

---

[14]Note that PlaSMA extends JADE. Thus, each static variable is only visible to the JVM. In distributed simulations on multiple machines, each machine requires its own static routing algorithm.

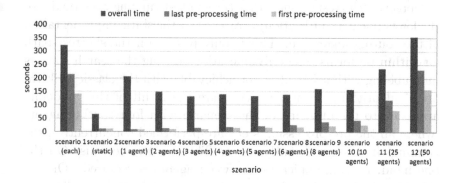

Figure 4.11: The average simulation time and the average time required
to create the hub labels (of 10 runs) of the different scenarios.

agents, but must be performed sequentially. Thus, it is reasonable
to outsource the cost-intensive operations to routing agents. As a
result, the pre-processing time as well as the total simulation time is
reduced. As mentioned above, in JADE each agent has its own process.
However, if there is just a single routing agent, the hardware utilization
is only about 25% on a quad-core processor, because all search queries
are executed by a single agent on a single core. Consequently, the
pre-processing time and the total simulation time can be reduced by
physically concurrently running routing agents as long as the number
of routing agents is lower or equal to the number of available cores. To
this end, the search queries are answered in parallel and the hardware
utilization is increased. If the number of routing agents exceeds the
amount of cores, the pre-processing for the creation of the hub labels
cannot be performed in parallel anymore and the total time for pre-
processing and simulation is increased. If each consumer agent has
its own routing agent, the performance is even higher compared with
each agent having its own algorithm. This is explained by the fact
that the message transfer within the multiagent system consumes
additional time.

Moreover, the results show that the shortest running time (and also the lowest memory utilization) is reached by the static implementation of the routing algorithm. This is only possible if the shortest-path algorithm performs read-only operations during the search such as hub labeling with a CH. It is not achievable with the classic Dijkstra or A* algorithm. Even if we compare the result of this modeling approach to the outcome of the scenario with four routing agents (where all the pre-processing steps are performed concurrently), the running time is significantly lower. This has two reasons: On the one hand, no communication between agents is required. On the other hand, hardware utilization is higher in the simulation because the algorithm can answer four requests concurrently at any time. When a computation is finished, the next shortest path is computed immediately afterwards. In contrast, with four routing agents each agent computes its assigned shortest-path requests. When an agent has finished its computation, it has to wait until the last agent has also finished its computations in order to ensure the correctness of the simulation which implements a *conservative synchronization mechanism* (cf. Section 6.4.1). If the agents continued to perform tasks of the next time slice, the already processing agent might receive messages from the future and the consistency of the simulation is not ensured. Thus, the next time slice is only started when all agents have finished their tasks of the current time slice.

Although it is also possible to increase the scalability of simulation platforms (Ahlbrecht, Dix, Köster, Kraus, and Müller, 2014) or adding some optimization support for executing simulations on parallel processors (Sano, Kadono, and Fukuta, 2014), shortest-path searches remain one of the most cost-intensive operations of the agents in simulated logistics scenarios. In general, this is especially true for multiagent systems which are applied in industrial processes. The results reveal that a slight technical restriction of the agents' autonomy by applying a single algorithm saved in a static variable (which is part of all the agents) clearly results in the lowest runtime of the simulation and the lowest memory consumption.

As long as all agents run on the same machine (and the same JVM), the disadvantage of less autonomy in this modeling approach is theoretical rather than practically relevant. For instance, the robustness could even be guaranteed by a second redundant static instance of the algorithm. Privacy is also guaranteed, because the agents must not reveal their search queries to any other agent.

However, if the *full* autonomy of the agents must be guaranteed, another option is to create several routing agents that receive routing requests, perform shortest-path searches, and provide the results. Although this approach consumes more time, i.e., because of the increased time consumption for message transfer and synchronization of agents, it can still profit from concurrent calculations as long as the number of routing agents is approximately equal to or lower than the number of available cores. Otherwise, the redundant algorithms consume a high amount of memory (in particular if the shortest-path searches are performed on large graphs) as well as time for communication and computation. In addition, shortest paths are not performed physically concurrently. In the extreme case it is even preferable that each consumer agent has its own algorithm. In this case the autonomy of the agents is maximized and less communication is required.

In conclusion, applying a static hub labeling algorithm in MAS for transport logistics, which is part of all agents, allows for concurrent calculations, improves the runtime performance of the simulation significantly, and reduces the memory usage. In contrast to the other established modeling approaches, this allows the application of MAS in real-world logistics scenarios on large infrastructures with less hardware requirements.

4.6 Summary and Conclusion

This section presented a multiagent system (MAS) for the optimization of transport logistics. Beside the consideration of domain-dependent requirements, the MAS allows splitting up the overall problem to multiple small problems to handle the high complexity. In addition, it provides a flexible and robust behavior, which is especially relevant in dynamic environments.

The implemented pre-processing (a clustering) is applied to accelerate the search. It allows solving also large problems in a short time. The pre-processing computes a general static solution which is further improved and refined by the MAS.

The developed communication and negotiation mechanisms are designed to support a high degree of concurrent negotiations and computations which can be performed, e.g., on a cloud computing infrastructure. In addition, the mechanisms ensure that the decision-making authority remains at the selfishly acting agent of the represented object/entity. The communication mechanisms are stable which means that even in competitive situations it is not possible to manipulate the outcome of a negotiation for selfish purposes. In scenarios with multiple competing market participants or scenarios where confidential data has to be kept private, stable communication mechanisms are particularly relevant.

The most time- and resource-intensive operations of the MAS are the decision-making processes of the vehicle agents. As the agent's decisions are predicated on solving individualized and constrained routing problems, they require time and memory-efficient routing algorithms. The algorithms are designed to directly consider domain-dependent requirements within the search process such as time windows, individual start and end locations of vehicles, capacity constraints, and priorities of orders. Moreover, they allow to solve problems with one or multiple depots as well as problems with orders which have to be transported directly from the pickup to the delivery location. A branch-and-bound algorithm was developed to solve small problems optimally. The algorithm applies efficient bit-vector operations in $O(1)$ for the constraint checks at each search node which allows verifying millions of nodes in a few milliseconds without allocating additional memory during the search. The optimality of this algorithm is proven. In addition, a novel approach based on Nested Rollout Policy Adaptation (NRPA) was developed which computes state-of-the-art solutions for medium and large-size problems with pickups and deliveries. The structure of the MAS allows saving, extending, and reusing a once computed existing policy. Thus, the algorithm enables the agents to retain their generated knowledge. In general, other domain-dependent requirements and changing objective functions can simply be considered by an adaptation of the decision-making process of the vehicle agents. Therefore, only the *rollout* function of the NRPA algorithm or the constraint check and the cost function of the branch-and-bound algorithm has to be changed.

Finally, the chapter focused on the impact of shortest-path searches in multiagent-based transport processes and investigated the implementation of different modeling approaches and time- and memory-efficient autonomous transport processes. It revealed that established shortest-path algorithms, which are in P, have a higher impact on the overall runtime than the solver for the NP-hard traveling salesman problem. The implementation of the state-of-the-art hub-labeling with contraction hierarchies algorithm reduces the runtime significantly. The runtime is further decreased by a suitable agent model, e.g., if shortest-path searches are answered by routing agents or only a single static implementation of the shortest-path algorithm is implemented. Although shortest-path searches are completely neglected by all other authors who have developed MASs for transport logistics which are presented in Section 3.4, they are absolutely relevant for the real-world application of MASs.

5 Benchmark Evaluation

In order to precisely validate the performance of the implemented multiagent system (MAS) quantitatively, this chapter evaluates the system on benchmark problems. The most established and often applied benchmarks for Vehicle Routing Problems (VRP) are the Solomon benchmark (Solomon, 1987) as well as the benchmark of Homberger and Gehring (2005). Consequently, this chapter focuses on these two benchmarks.

Firstly, Section 5.1 discusses the difficulties of benchmarks for dynamic VRPs. Next, Section 5.2 presents the benchmark sets of Solomon (1987) and Homberger and Gehring (2005). Section 5.3 provides a parameter investigation to determine a suitable configuration for different settings. Based on the knowledge gained, Sections 5.4–5.5 evaluate the dispAgent approach with the benchmark sets of Solomon (1987) and Homberger and Gehring (2005), respectively. Finally, Section 5.6 concludes the results.

5.1 Benchmarks for Dynamic VRPs

As mentioned by Pillac et al. (2013, p. 8) and van Lon and Holvoet (2013, p. 48), there is no standard reference benchmark for solving dynamic VRPs. This might be caused by the fact that a dynamic benchmark requires a unified platform and a standardized model to simulate system-specific dynamic dependencies and interactions over time. The platform has to ensure the reproducibility of simulated stochastic events. In order to allow pertinent and consistent comparison, a unified time-model must be used. For example, if such a platform provides a discrete-time simulation, the granularity and length of time slices has to be standardized to ensure that all information (such as changing positions of entities and other events) is available at the same time and no approach is privileged by having earlier or more fine-granular access to dynamic information. Therefore, the platform is responsible for synchronizing all events at every point in time. For instance, all positions of vehicles must be updated in each time slice to have a consistent system state. In addition, reference scenarios must specify in detail which dynamic event occurs at which point in time. It is necessary that the underlying traffic infrastructure allows updating and defining the position of each entity at any point in time. The difficulty is that such a platform has to support all kinds of approaches. Chapters 2–3 give some insight into the wide variety of different solution methods ranging from classical OR-approaches such as dynamic programming and meta-heuristics like tabu-search to multiagent systems.

In conclusion, dynamic well-established benchmarks may not exist because there is no standardized platform nor are there standardized logistics scenarios which would allow comparing the high diversity of approaches with relatively little effort by simulation. Instead, authors of dynamic approaches such as Gendreau et al. (1999), Bent and Hentenryck (2004), and Chen and Xu (2006) evaluate their systems on problems which are customized adaptations of the problems provided by Solomon (1987) and Homberger and Gehring (2005). Thus, a significant comparison of the different approaches is not possible.

In order to evaluate multiagent systems in dynamic logistics transport scenarios, the PlaSMA simulation platform has been developed by the University of Bremen within the Collaborative Research Center 637 (SFB 637). The platform, which is presented in Section 6.4.1, was applied to evaluate the dispAgent approach in dynamic real-world scenarios of Hellmann Worldwide Logistics (cf. Chapter 6). Nevertheless, there remains the need to precisely investigate and compare the system performance to the best-known solutions of state-of-the art approaches. Therefore, this chapter focuses on the well-established and often applied static benchmark sets of Solomon (1987) and Homberger and Gehring (2005).

5.2 The Benchmarks of Solomon and of Homberger and Gehring

The benchmarks of Solomon (1987) and Homberger and Gehring (2005) define six types of problems:

- Sets $R1$ and $R2$ include problems with randomized customer locations.

- Sets $C1$ and $C2$ include problems with customers grouped in clusters.

- Sets $RC1$ and $RC2$ include problems with both clustered and randomized customer locations.

The time windows of sets $R1$, $C1$, and $R1$ are shorter and more restricted than those of sets $R2$, $C2$, and $RC2$. Thus, more vehicles are required to transport all orders in the restricted problems. In all problems, vehicles have limited capacities. There is no underlying transport infrastructure. Distances are computed by Pythagoras' theorem. It is assumed that driving a distance of 1 consumes 1 time unit. Solomon provides 56 small problem instances with up to 100 customers. Homberger and Gehring (2005) extend Solomon's problem set for larger size problems and provide 300 instances with 200 up to 1000 customers. The problems are based on the Solomon benchmark set. They provide 10 instances for each problem set and size. The goal is to satisfy all customer requests, minimize the total number of required vehicles with highest priority, and minimize the total distance with second highest priority.

Same as Kalina and Vokřínek (2012b, p. 1561), this investigation measures the result quality by comparing it to the best-known solution by means of the following equation:

$$quality = \frac{best\text{-}known\ solution}{measured\ solution} \times 100\ [\%]\,. \qquad (5.1)$$

Since the problem has multiple hierarchical objectives, the quality of reaching the best-known number of vehicles (VQ) as well as the shortest-known distance (DQ) is measured.

Both benchmarks are designed for large VRPs. Since the problems are NP-hard there exists (as yet) no optimal solver which determines the optimal solution of these problems. The best-known solutions of both benchmark sets are continuously updated on the SINTEF homepage.[1] The difficulty of solving these problems is further indicated by the fact that authors have continuously found improved solutions. For instance, for single problems best-known results were recently published on 15 September 2014 by Nalepa, Blocho, and Czech (2014).[2] This clearly substantiates that it is quite hard to determine best-known solutions, even for static solvers which are especially designed for solving these static VRPs with time windows and capacities. Therefore, no solver exists which could determine the best-known solutions in all problems. This is simply demonstrated as follows: Nalepa et al. (2014) published some new best-known solutions on 15 September 2014, but failed to determine the best-known solutions in other problems. For instance in problem $R\_1\_4\_2$ of Homberger's benchmark set, the shortest-known distance is $7,686.38$[3], while the solution's distance of Nalepa et al. (2014, p. 198) is $9,026.92$. Related to Equation 5.1, the subset (R_1) of the benchmark set has a distance quality (DQ) of 85%. Note that the changing solution quality of the best-known solution affects the comparison of approaches by Equation 5.1.

Furthermore, the artificial benchmark sets are abnormally restricted compared to real-world problems. For the computation of correct best-known solutions it is crucial to compute the Euclidian distances with double precision arithmetic. Using any lower precision arithmetic

---

[1] http://www.sintef.no/Projectweb/TOP/VRPTW/ (cited: 1.9.15).

[2] http://www.sintef.no/Projectweb/TOP/VRPTW/Homberger-benchmark/800-customers/ (cited: 1.9.15).

[3] http://www.sintef.no/Projectweb/TOP/VRPTW/Homberger-benchmark/400-customers/ (cited: 1.9.15).

will result in incorrect solutions.[4] Even rounding the distances at a fixed position after the decimal point might mean that a vehicle cannot reach the next stop on time, which is necessary to find the optimal solution. Obviously, in real-world processes such small delays of less than a nanosecond are absolutely insignificant.

5.3 Parameter Investigations

In order to find a suitable configuration of the multiagent system, this section analyzes the impact of varying the following parameters:

- the configuration of the NRPA algorithm (the number of iterations and the maximum number of performed rollouts),

- the probability of an order agent accepting the second best proposal instead of the best one to avoid getting stuck in local optima, and

- the runtime of the system.

Unless specified otherwise, the following investigations focus on a low constraint problem in which customers are grouped in clusters (C201) as well as a constraint problem with randomized customer locations (R101). Both problems are included in the benchmark set of Solomon (1987). In all runs, the maximum available number of couriers refers to the best-known solution.[5] As the algorithms contain components which use randomized variables, all experiments were repeated 10 times. Each parameter investigation is either performed on an Intel(R) Core(TM) i7-2620M CPU at 2.70 GHz or on an Intel(R) Core(TM) i7-5930K CPU at 3.50 GHz.

---

[4]http://www.sintef.no/Projectweb/TOP/VRPTW/Solomon-benchmark/
100-customers/ (cited: 1.9.15).
[5]Taken from http://www.sintef.no/Projectweb/TOP/VRPTW/
Solomon-benchmark/100-customers/ on the 1.3.2015.

5.3.1 Configuration of the Decision-Making Algorithms

The optimal branch-and-bound algorithm is applied to solve problems with less than 12 stops. Therefore, a maximum of 39,916,800 different stop combinations are possible (cf. Chapter 2). However, multiple sub-trees can be pruned because of constraints such as time windows, capacity, as well as the pickup before delivery constraint. Finally, tests reveal that in most cases less than about 1,000,000 expansions are required to determine the optimal solution. In order to avoid the decision-making process being delayed in rare unconstrained cases, the algorithm stops after performing 2,000,000 expansions at the most. The anytime behavior of the algorithm ensures that it returns the best-found solution. A more detailed evaluation of the branch-and-bound algorithm is provided in Section 4.4.1 as well as by Edelkamp and Gath (2013).

Although suitable parameters of the NRPA algorithm for solving TSPs are investigated in detail in Section 4.4.2, it may be expedient to reduce the runtime (and solution-quality) of the agents' decision-making process in the multiagent-based approach for solving VRPs. Shorter decision-making of the agents allows for more negotiations during the same runtime. Therefore, different limits, ranging from 100 up to 100,000 for the maximum number of rollouts, were investigated.[6] Each configuration runs 10 times on the R101 as well as on the C201 problem. All other parameters, e.g., the runtime of 15 minutes, remain constant.

Figures 5.1–5.2 depict the average number of unserved orders for each experiment while Figures 5.3–5.4 show box-whisker plots for the total distances driven in these experiments. Moreover, Figure 5.5 to 5.8 examine the number of performed negotiations for finding a vehicle if the order is unallocated (cf. Section 4.3) as well as the number of performed dynamic negotiations for improving the allocations by changing the vehicle if the order is allocated (cf. Section 4.3.1).

---

[6]If the maximum number of rollouts is performed, the algorithm stops and the best-found result is returned (cf. Section 4.4.2).

Section 4.4.2 shows that best results for solving a single TSPTW from scratch by NRPA are computed with a maximum number of performed rollouts of about 40,000. Nevertheless, the results visualized in Figure 5.1 to 5.7 prove that a much lower number of rollouts clearly leads to more efficient solutions for the VRP in the multiagent-based approach, although this decreases the solution quality of single TSPs. This is explained by a significant increase in the number of performed agent negotiations, because the most cost-intensive part of the agents' decision-making processes (solving TSPs) requires substantially less time. As a result, the multiagent-based negotiations are more relevant for the optimization than an optimal (or near optimal) solution for the TSP-like sub-problems. For instance, if a vehicle agent sends an unacceptable offer to an order agent, this offer might be improved in a consecutive negotiation, because the solver might find a more efficient tour.

In general, there is a trade-off between exploration and exploitation in the NRPA algorithm, which can be adjusted by the parameters *level* and *iterations*. More iterations result in higher exploration of the search space. If the algorithm is stopped before all $iterations^{level}$ rollouts are performed, higher levels have no effect (cf. Section 4.4.2). Since the runs were performed with 64 iterations, it seems that in most experiments the NRPA search includes 2 or 3 levels only. However, the agents learn and improve their policy. As the TSPs mostly vary in some stops only, each vehicle agent maintains a policy which is updated with the recently improved policy after each decision-making process. The next NRPA search is initialized with this policy (cf. Section 4.4.2). Nevertheless, a minimum number of rollouts is required to improve the policy and adapt it to the current situation. A suitable number of rollouts depends on the size of the problem. While in the R101 problem 100 orders are allocated to 19 vehicles, in the C201 problem 100 orders must be allocated to merely 4 vehicles. Thus, also the complexity of solving the sub-problems increases in the C201 problem. Subsequently, more rollouts should be applied. Figure 5.1 to 5.4 show that a suitable limit for both problems is between 2,000 and 10,000 negotiations. Higher maximum numbers of rollouts particularly affect the more relevant objective function of minimizing the number of unserved orders. Furthermore, it has a minor influence on distance minimization.

In the next experiments, the effects of varying the numbers of iterations are investigated. Keep in mind that the agents initialize the policy with established results of recent decision-making processes. All other parameters remain constant. Figures 5.9–5.10 visualize the result quality of the R101 problem, while Figures 5.11–5.12 show the result quality of the C201 problem.

The results show that the NRPA algorithm should perform 64 iterations to reach an adequate exploration rate. While the solution quality of the R101 problem (which requires less complex decisions) only increases slightly, the solution quality of the C201 problem (which requires more complex decisions) rises significantly with 64 iterations. If the number of iterations further increases, the exploration avoids the exploitation of promising results which leads to a lower overall solution quality in both problems. Note that the maximum number of rollouts is restricted to 10,000 in these experiments. It is obvious that in general a larger number of iterations would yield more efficient solutions if the search were not stopped after a maximum number of rollouts have been performed. However, the goal of this experiments is to find the best configuration if the time for decision-making is limited in order to allow a high amount of negotiations. The number of iterations has a minor effect on the second objective function of minimizing the total distance.

5.3.2 Avoiding Local Optima

In order to avoid getting stuck in local optima, order agents can accept the second best proposal instead of the best one if their transport request has not been satisfied (cf. Section 4.3). The following experiments investigate how often the second best proposal should be accepted and focus on the impact this has on the overall solution quality. Therefore, several experiments with different probabilities to accept the second best proposal were performed. Each experiment

runs 10 times. Figures 5.13–5.14 show the results of the R101 problem.
Figures 5.15–5.16 depict the results of the C201 problem. The figures
prove that in the investigated problems the strategy indeed increases
the solution quality significantly. The best results in both problems
are computed if the order agents accept the second best proposal
instead of the best one in 10 % of negotiations. If the best proposal
is always accepted, the order agent could get caught up in a loop of
negotiations. In order to avoid this loop, the order agents rarely accept
the second best proposal, which allows for negotiations with other
participants. Nevertheless, later on another vehicle can satisfy the
order's request at lower cost and without rejecting another transport
request. In this case, the order agent would identify this vehicle in
the dynamically ongoing negotiations (cf. Section 4.3.1) and would
decide to switch the transport service provider. In turn, the results
show that accepting the second best proposal in more than 10% of
negotiations decreases the solution quality in both problems.

5.3.3 Runtime Investigations

The more time there is, the more multiagent negotiations can be
performed and the more the solution quality increases. The next
experiments focus on the trade-off between the runtime and the
solution quality. Therefore, the R101 problem was solved in three
experiments. The experiments only differ in the runtime. Each
experiment runs 10 times. Figures 5.17–5.18 visualize the results.
They show that the system continuously improves the solution the
longer it keeps running. However, this only affects the higher priority
objective function of minimizing the number of unserved orders. The
total distance remains approximately constant. Furthermore, the
solution quality increases monotonically and slightly. Thus, the best
strategy for real-world application is to keep the system running
until the result is required. In general, the multiagent system should
never be stopped or run until all operations are finished, because it is
especially designed for dynamic environments and can consider also
unexpected events during operations.

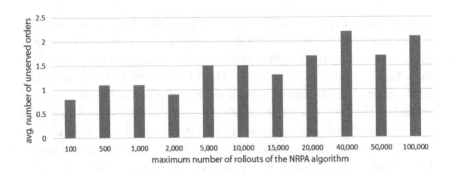

Figure 5.1: Average number of unserved orders of ten runs in eleven experiments. The experiments only vary in the upper bound for the number of rollouts of the NRPA algorithm. In each experiment, the R101 problem is solved. The amount of available couriers is set to the best-known solution.

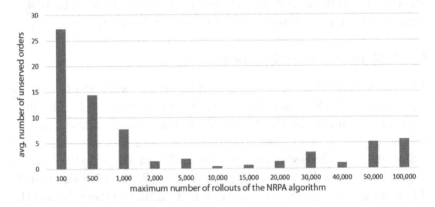

Figure 5.2: Average number of unserved orders of ten runs in twelve experiments. The experiments only vary in the upper bound for the number of rollouts of the NRPA algorithm. In each experiment, the C201 problem is solved. The amount of available couriers is set to the best-known solution.

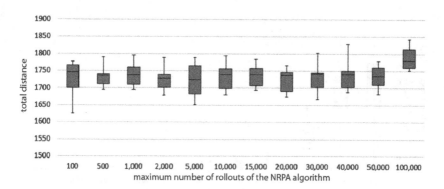

Figure 5.3: The box-whisker plots include the total (Euclidian) distances of the experiments illustrated by Figure 5.1. Note that the total distance might be lower than in the best-known solution. In this case, the solution includes unserved orders.

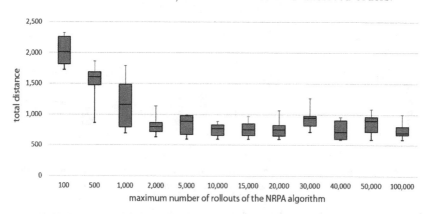

Figure 5.4: The box-whisker plots include the total (Euclidian) distances of the experiments illustrated by Figure 5.2. Note that the total distance might be lower than in the best-known solution. In this case, the solution includes unserved orders

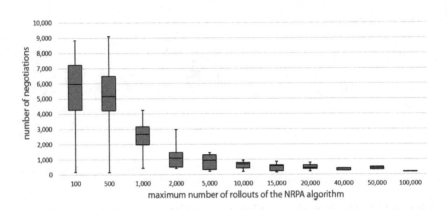

Figure 5.5: The box-whisker plots include the total number of performed
negotiations for finding a vehicle (cf. Section 4.3) in the
experiments which are illustrated by Figure 5.1.

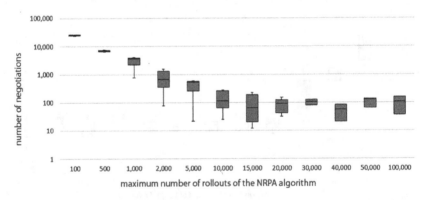

Figure 5.6: The box-whisker plots include the total number of performed
negotiations for finding a vehicle (cf. Section 4.3) in the
experiments which are illustrated by Figure 5.2. Note that
the scaling of the y-axis is logarithmic.

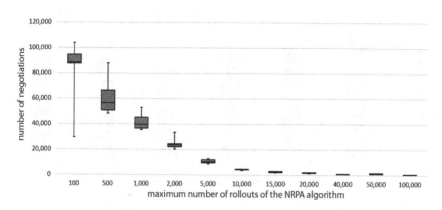

Figure 5.7: The box-whisker plots include the total number of performed
dynamic negotiations (cf. Section 4.3.1) for improving the
solution quality in the experiments which are illustrated by
Figure 5.1.

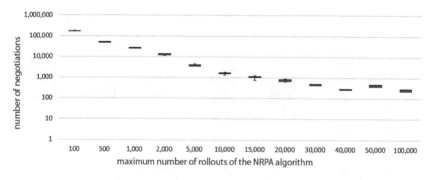

Figure 5.8: The box-whisker plots include the total number of performed
dynamic negotiations (cf. Section 4.3.1) in the experiments
which are illustrated by Figure 5.2. Note that the scaling of
the y-axis is logarithmic.

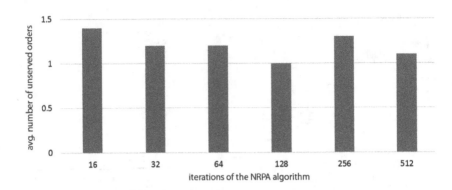

Figure 5.9: Average number of unserved orders of ten runs in six experiments. The experiments only vary in the upper bound for the number of rollouts of the NRPA algorithm. In each experiment, the R101 problem is solved. The amount of available couriers is set to the best-known solution.

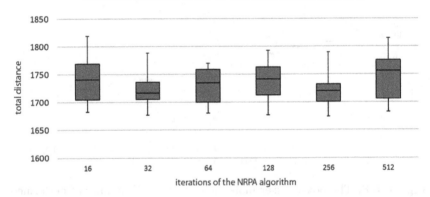

Figure 5.10: The box-whisker plots include the total (Euclidian) distances of the experiments illustrated by Figure 5.9. Note that the total distance might be lower than in the best-known solution. In this case, the solution includes unserved orders.

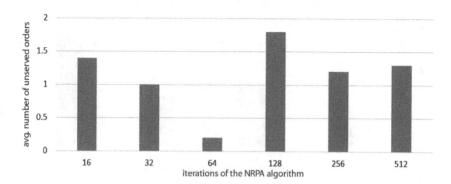

Figure 5.11: Average number of unserved orders of ten runs in six experiments. The experiments only vary in the upper bound
for the number of rollouts of the NRPA algorithm. In each
experiment, the C201 problem is solved. The amount of
available couriers is set to the best-known solution.

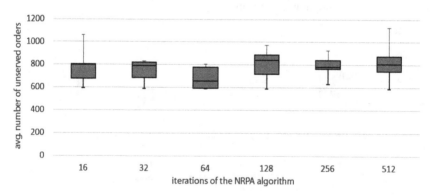

Figure 5.12: The box-whisker plots include the total (Euclidian) distances of the experiments illustrated by Figure 5.11. Note
that the total distance might be lower than in the best-
known solution. In this case, the solution includes unserved
orders.

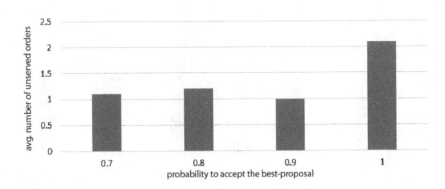

Figure 5.13: Average number of unserved orders of ten runs in four experiments. The experiments only vary in the probability of
an order agent accepting the second best proposal instead
of the best one. In each experiment, the R101 problem
is solved. The amount of available couriers is set to the
best-known solution.

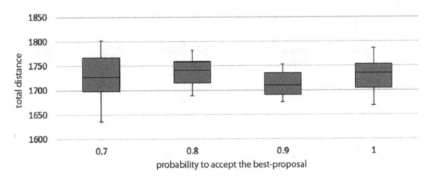

Figure 5.14: The box-whisker plots include the total (Euclidian) distances of the experiments illustrated by Figure 5.13. Note
that the total distance might be lower than in the best-
known solution. In this case, the solution includes unserved
orders.

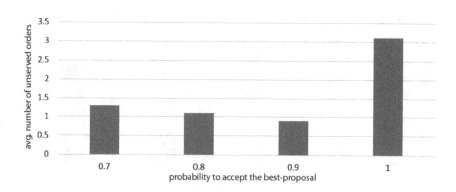

Figure 5.15: Average number of unserved orders of ten runs in four experiments. The experiments only vary in the probability of an order agent accepting the second best proposal instead of the best one. In each experiment, the C201 problem is solved. The amount of available couriers is set to the best-known solution.

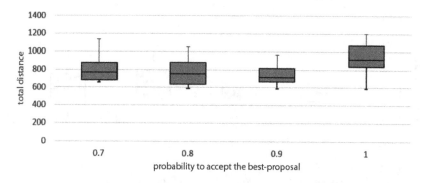

Figure 5.16: The box-whisker plots include the total (Euclidian) distances of the experiments illustrated by Figure 5.15. Note that the total distance might be lower than in the best-known solution. In this case, the solution includes unserved orders.

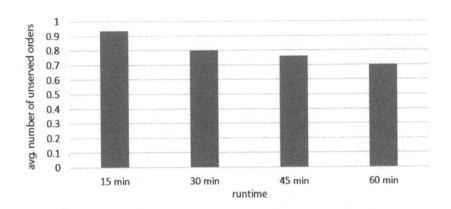

Figure 5.17: Average number of unserved orders of ten runs in three experiments. The experiments only vary in the running time. In each experiment, the R101 problem is solved. The amount of available couriers is set to the best-known solution.

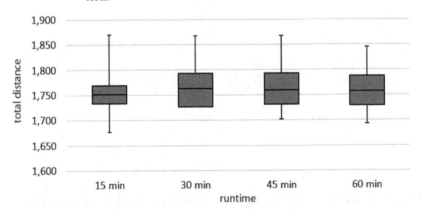

Figure 5.18: The box-whisker plots include the total (Euclidian) distances of the experiments illustrated by Figure 5.17.

5.3.4 Conclusion and Discussion

In conclusion, complex TSP-like sub-problems must be solved in a short time, although this might result in a lower solution quality of the sub-problems. It increases the overall solution quality because more multiagent negotiations can be performed in the same space of time. It turns out that a suitable configuration of the NRPA algorithm is to limit the rollouts between 2,000 and 10,000 and to perform 64 iterations for solving the less constraint clustered problem as well as the constraint randomized problem. Although searches are initialized with memorized policies, the values should not be defined below these limits to ensure sufficient exploration. Higher values for the number of iterations limit the exploitation and more rollouts lead to less multiagent negotiations.

In addition, the order agents should rarely accept the second best proposal instead of the best one, to guide the search into a new direction and to increase the exploration rate. A suitable strategy is to accept the second best proposal in 10% of all negotiations if the order's transport request is unsatisfied.

Furthermore, the results show that the solution quality can vary because of randomized components of the approach. To reduce this effect in real-world applications, the runtime can be increased which subsequently improves the overall solution quality. For example, in dynamic environments the multiagent system should never be stopped anyway, as it continues optimizing tours during operation.

5.4 Solomon Benchmark

This section investigates the performance with the benchmark of Solomon (1987).

5.4.1 Experimental Setup

The evaluation was performed on an Intel(R) Core(TM) i7-5930k CPU at 3.5 GHz. The computer was equipped with 32 GB RAM and with Windows 7 64bit. The first set of experiments was stopped after 15 minutes, the second one after 60 minutes. Both sets contain all problem instances of Solomon's benchmark set with 100 customers. Since the implementation used did not automatically add new vehicles, firstly all experiments were performed with the best-known number of vehicles. Next, each instance whose result contained unserved orders was started again with an additional vehicle. This procedure was continued until a solution with zero unserved orders was found. The NRPA algorithm performed 64 iterations and was stopped after 5,000 rollouts (keep in mind that each agent saves it's learned policy and initialized the next run with this policy). In 10% of all initial negotiations, the order agent accepted the second best offer instead of the best one.

5.4.2 Results

In all problem instances of Solomon's benchmark, the dispAgent software system reaches a total VQ of 91.42% (after 15 minutes) and of 93.53% (after 60 minutes) on average.[7] In about 46.43% (after 15 minutes) and 51.79% (after 60 minutes) of all solutions, the best-known result in the VQ has been established. All problems have an average DQ of about 72.48% (after 15 minutes) and of about 76.30% (after 60 minutes). Note that the number of instances included in each problem type varies (cf. Section 5.2).

Table. A.1 shows the VQ and DQ separately for each problem type. These experiments were stopped after 15 minutes. Table. 5.2 shows the VQ and DQ separately for each problem type. These experiments were stopped after 60 minutes. The detailed results for each problem instance are provided in Appendix A. In addition, the evaluation of some instances was also performed on the parallel supercomputing system of the HLRN on a node containing two 12-core Xeon IvyBridge CPUs, E5-2695 v2 @2.4GHz and 64GB RAM. The manually observed CPU usages when solving Solomon's R101 problem instance (contain-

---

[7]All best-known solutions in this investigation are taken from http://www. sintef.no/Projectweb/TOP/VRPTW/ on the 1.3.2015).

ing 19 vehicles and 100 orders) show, that the dispAgent system indeed supports a high level of parallel computations. For instance, while solving a problem with 19 vehicles (whose responsible agents must perform the most CPU-intensive operations), the UNIX *top* command shows that the CPU usage is more than 13,000%, which indicates that at least 13 processors reach their full capacity utilization. The same behavior was observed when the system was run in the cloud in real-world operations by the tiramizoo GmbH (cf. Section 7). In these runs doubling the number of cores cut the running time in half.

Table 5.1: The VQ and DQ separated by problem types of the Solomon benchmark set with 100 orders. Each experiment was stopped after 15 minutes.

Type	VQ	DQ
C1	100.00 %	78.67%
C2	88.88%	52.74%
R1	92.26 %	87.73%
R2	75.00 %	61.45%
RC1	92.93%	92.04%
RC2	81.25%	60.24%
All	91.42%	72.48%

Table 5.2: The VQ and DQ separated by problem types of the Solomon benchmark set with 100 orders. Each experiments was stopped after 60 minutes.

Type	VQ	DQ
C1	98.90%	73.04%
C2	85.71%	63.33%
R1	93.46%	89.13%
R2	83.33%	65.61%
RC1	95.83%	93.64%
RC2	89.65%	67.70%
All	93.53%	76.30%

5.4.3 Discussion

The experiments verify the observations presented in Section 5.3.3 that longer runtime leads to increasing tour quality. In addition, it was shown that increasing the number of cores improves the result quality and reduces the runtime.

Small deviations in the solution quality of single problem instances are reduced to randomized components such as the NRPA algorithm. Increasing the number of runs and taking average values would eliminate this randomness. However, also all other authors whose approaches are discussed in Section 3.4 did not perform such an extensive investigation, because the gained knowledge hardly justifies the excessive hardware utilization which is required due to the high complexity of the problem instances.

In general, the results indicate that the multiagent system achieves a better result quality in solving the restricted problem sets (R1, C1, RC1) than in solving the unrestricted problem sets (R2, C2, RC2). This is because the constraints are directly considered in the agents' decision-making process (cf. Section 4.4). The algorithms are optimized and developed for real-world restricted problems. For instance, the optimal algorithm presented in Section 4.4.1 decreases the search space by cutting constraint-violating sub-trees. As long as no constraints are violated, the search space is not reduced. Also the tour evaluation (the result of the *rollout* function of the NRPA solver (cf. Algorithm 3)) is clearer in restricted problems, because of the high "penalty" for constraint violations. Thus, for solving less restricted problems in which 100 orders are transported by only 3 vehicles, the application of a standard TSP solver for the agent's decision-making might improve the result quality in these artificial benchmark instances.

The result qualities outperform those of Kalina and Vokřínek (2012a) (which have depending on the applied strategy a VQ of 83.5% and

87.9% as well as a DQ of 49.4% and 66%),[8] and are competitive with those of Leong and Liu (2006) (having an overall VQ of 92.3% and DQ of 96.4 %).[9]

Note that both solutions of Kalina and Vokřínek (2012a) and Leong and Liu (2006) are rather centralized implementations of decentralized algorithms instead of real multiagent systems as they are defined in Chapter 3. Their agents are objects in object-oriented programing languages rather than intelligent agents.[10] In both MASs agents do not act selfishly. In addition, they neither use an FIPA compliant agent management system nor a message transfer system. On the one hand, they cannot profit from the advantages of a decentralized MAS such as increased robustness or the consideration of privacy aspects. On the other hand, they have lower run-times or rather more time for optimization, because no overhead is required for sending messages etc. Especially benchmark evaluation benefits from this (from a multiagent-based point of view) lack of modeling.

Fischer et al. (1996) also evaluated their MAS on the 12 randomized $R1$ problem instances. As they present their solutions of these problems in detail, the result quality can be updated by considering the current best-known solutions. If the solution quality is compared to that of the dispAgent system, the dispAgent system clearly outperforms the approach of Fischer et al. (1996). While Fischer et al. (1996) compute solutions with an average VQ of 78.11% and an average DQ

---

[8]The results of the Algorithm-DI and algorithm-DIA of Kalina and Vokřínek (2012a) are neglected, because the authors performed three different runs with varying configurations/strategies and finally took only the best result ("With respect to the *dynamicImprove* function, we tested all three presented negotiation methods in the applicable [...] settings. The presented results correspond to the best of these three runs.") (Kalina and Vokřínek, 2012a)[p. 6].

[9]For the correctness: These quality measures are indicated by Kalina and Vokřínek (2012a) and refer to the best-known solutions on the date of publication (the computations of VQ and DQ are based on the best-known solutions; cf. Eq. 5.1). Today (1.9.2015), better solutions for some tours have been found, which slightly decreases the solution quality (but insignificantly).

[10]The interested reader is referred to Wooldridge (2009, p. 28 - 30) who discusses the differences between objects of object-oriented programming languages and agents.

of 81.97%, the dispAgent solutions have an average VQ of 92.26% (after 15 minutes) and of 93.46% (after 60 minutes) as well as an average DQ of 87.73% (after 15 minutes) and of 89.13% (after 60 minutes) on this subset of the benchmark. Note that the results of Fischer et al. (1996) were already published in 1996.

5.5 Homberger and Gehring Benchmark

This section investigates the performance with the benchmark of Homberger and Gehring (2005).

5.5.1 Experimental Setup

The evaluation was performed on an Intel(R) Core(TM) i7-5930k CPU at 3.5 GHz. The computer was equipped with 32 GB RAM and with Windows 7 64bit. All experiments were stopped after 60 minutes. The problem set contains all problem instances of the Homberger and Gehring Benchmark set with 200 customers. The configuration is the same as in the experiments with the Solomon Benchmark (cf. Section 5.4.1).

5.5.2 Results

In the benchmark set of Homberger and Gehring (2005) with 200 orders, the computed solutions have an average VQ of 93.78% and an average DQ of 60.86%.[11] In 40.00% of all solutions, the best-known result in the VQ is obtained. Table. 5.3 shows the VQ and DQ of each problem type.

5.5.3 Discussion

Similar to the Solomon benchmark, the difference in the result quality of restricted compared to unrestricted problems can be observed (cf.

---

[11]All best-known solutions in this investigation are taken from http://www. sintef.no/Projectweb/TOP/VRPTW/ on the 1.3.2015).

Table 5.3: The VQ and DQ separated by problem types of the Homberger and Gehring benchmark set with 200 orders.

Type	VQ	DQ
C1	97.93%	75.34%
C2	85.71%	46.47%
R1	98.38%	70.07%
R2	83.33%	50.18%
RC1	94.74%	69.25%
RC2	79.63%	56.59%
All	93.78%	60.86%

Section 5.4.3). Also in terms of quality the results are competitive with or even outperform those of Kalina and Vokřínek (2012a) (which have depending on the applied strategy a VQ of 92.4% and 94.5% as well as a DQ of 24.6% and 46.6[12]).[13]

5.6 Summary and Conclusion

The goal of this chapter has been to precisely validate the performance of the dispAgent approach. Before starting the benchmark evaluation, the impact of relevant parameters as well as a suitable configuration of the MAS for different settings has been analyzed. Unfortunately, there is no adequate and established benchmark for dynamic problems, because there exist neither standardized platforms nor standardized logistics scenarios which allow comparing the highly diverse approaches

---

[12]The results of the Algorithm-DI and algorithm-DIA of Kalina and Vokřínek (2012a) are neglected, because the authors performed three different runs with varying configurations/strategies and finally took only the best result (Kalina and Vokřínek, 2012a)[p. 6].

[13]For the correctness: These quality measures are indicated by Kalina and Vokřínek (2012a) and refer to the best-known solutions on the date of publication (the computations of VQ and DQ are based on the best-known solutions; cf. Eq. 5.1). Today (1.8.2015), better solutions for some tours have been found, which slightly decreases the solution quality (but insignificantly).

with reasonable effort. Therefore, the result quality is compared to best-known solutions of the well-established and often applied static benchmark instances of Solomon (1987) and Homberger and Gehring (2005).[14]

Although the domain of static VRPs and in particular the approaches for solving the instances included in the benchmark sets of Solomon (1987) and Homberger and Gehring (2005) are clearly dominated by Operations Research (OR) methods, the dispAgent software system computes suitable solutions for these artificial problems. Especially for the most important objective function (the reduction of the number of vehicles), the best-known solution is often computed. In Solomon's benchmark, tours with the best-known number of vehicles are computed in 46.43% (after 15 minutes) and in 51.79% (after 60 minutes). In the benchmark of Homberger and Gehring with 200 customers tours with the best-known number of vehicles are computed in 40.00% of all experiments. Considering the high complexity of these artificial problems (cf. Section 5.2), this is remarkable because even the static OR solvers have problems finding best-known solutions although they are specifically developed to solve the benchmark instances and cannot be applied in dynamic real-world environments. If the solution quality of computed tours is compared to the quality of tours computed with other multiagent systems, which were evaluated with these benchmarks, the dispAgent approach is at least competitive with or even outperforms other multiagent systems.

---

[14]Nevertheless, the dispAgent was also evaluated in dynamic real-world scenarios (cf. Chapter 6).

6 A Case Study: Groupage Traffic

The dispAgent approach enables the computing of efficient solutions for complex problems in dynamic environments. This case study investigates decentralized, autonomous multiagent-based processes in fine-grained simulations for the support of planning and control in groupage traffic, based on real-world application at the Bremen office of Hellmann Worldwide Logistics GmbH & Co. KG. The case study shows that the dispAgent approach enhances complex and dynamic processes, e.g., in groupage traffic, with an adaptive, robust, reactive, and customized system behavior which increases the efficiency as well as the service quality by reliable deliveries.

Firstly, Section 6.1 starts with an overview of groupage traffic in general. Secondly, Section 6.2 introduces Hellmann Worldwide Logistics. After a brief look at the company history, the section focuses on the Bremen office and the *System Alliance* which is a cooperation network that has considerable influence on the processes in groupage traffic. Section 6.3 summarizes the most relevant results of a detailed documentation and analysis of the processes in groupage traffic at the Bremen office. It discloses the interdependencies between work sequences and actors. In addition, all feasible and available data which can be processed by an information system is collected. Moreover, the section identifies processes which reveal a high optimization potential. Next, Section 6.4 shows that the dispAgent approach optimizes planning and control in groupage traffic. Therefore, it presents a pertinent simulation of a dynamic real-world scenario at the Bremen office of Hellmann Worldwide Logistics. Finally, Section 6.5 concludes the chapter.

6.1 Groupage Traffic

In groupage traffic, several orders with less-than-truckload (LTL) freights are served by the same truck to decrease total costs. In pickup tours, trucks transport loads from their origin to a local depot where the goods are consolidated to build economical loads. Through LTL networks the load is transported to a depot in the destination area where a cooperating forwarding agency delivers each item to its final destination in onward carriage. Changing amounts and individual qualities of packages like weight, volume, priority, and value increase the complexity of process planning. The high degree of dynamics that result from unexpected events aggregates handling the complexity in real situations. In addition, the exact amount and properties of packages are not known in advance. Actual capacities are only revealed while serving tasks. Furthermore, undelivered loads in pre-carriage decrease the trucks' capacities in onward carriage. To react to changing traffic conditions and delays at incoming goods departments, it is essential to adapt tours and timetables and also to consider all relevant constraints.

The dynamics and complexity of planning and scheduling processes require an efficient, proactive, and reactive system behavior to improve the service quality while ensuring time- and cost-efficient transportation. The abstract problem of the dispatching process in pre- and onward carriage refers to the well-known Vehicle Routing Problem (VRP) (cf. Section 2.2). In general, the VRP is concerned with determining tours with minimum costs for a fleet of vehicles to satisfy customer requests at different destinations. The start and end point of each tour is the depot. Moreover, further constraints, e.g., time windows, capacities, as well as time consumption at the warehouse/customer have to be considered.

In forwarding agencies information technologies support dispatchers in their decision-making, but tours are still created manually on the basis of individual long term experience. Indeed, there are several professional transport management systems (TMS) for the planning and control of transport processes (a comprehensive overview of TMS

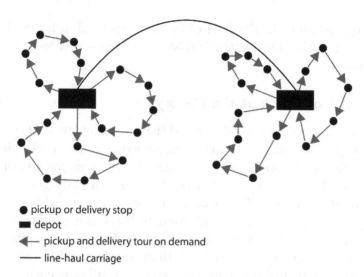

● pickup or delivery stop
■ depot
◀— pickup and delivery tour on demand
—— line-haul carriage

Figure 6.1: The main processes in groupage traffic (adapted from Cardeneo, 2008, p. 783).

applied in Germany is provided by Ten Hompel (2012, pp. 26 - 28)). However, the dynamics in logistics and customized requirements of the application domain are often neglected. Thus, the goal is to optimize the planning and control processes by providing operational plans, tours, and routes automatically while considering all relevant data to enable optimal decision-making at any point during operation.

6.2 Hellmann Worldwide Logistics

Founded in 1871, the owner-managed company Hellmann Worldwide Logistics GmbH & Ko. KG (hereinafter Hellmann) is one of the biggest logistics services providers in the world. In 2014, Hellmann achieved a sales volume of 3 Billion Euro with 12,872 employees in its 244 offices in 56 countries (Hellmann Worldwide Logistics GmbH & Co. KG, 2015). Products and services cover freight forwarding services by air, rail, and road, courier, express, and parcels services, as well as services in contract logistics. Moreover, they offer sector-specific industry solutions in the field of fashion, electronics, hospital and military logistics, automotive industry, and temperature-controlled transports.

6.2.1 A Brief Look at the History

Hellmann was founded in 1871 by Carl Heinrich Hellmann in Osnabrück, Germany. The company started transporting goods in Osnabrück and in the neighboring towns by horse-drawn carriages. As one of the first, Hellmann substituted the horse-drawn carriages by coal-fired vehicles in 1925. Ten years later, the company had more than 60 employees and started to consolidate heterogeneous goods in a single tour which was a vast innovation and increased the efficiency of transport significantly. In 1959, Hellmann founded the so-called *11er Kreis* with five other medium-size forwarding agencies. This was the first cargo cooperation network in Germany and laid the foundations for today's *System Alliance*. In the following years, there was an increasing demand for courier and express services. To be able to compete with other companies which had specialized in courier and express services, Hellmann and the *11er Kreis* founded Deutsche Paket Dienste (DPD). In 2001, DPD was sold to the French Post Office.

Motivated by globalization and by the rising need to open up new and attractive markets in Asia, Hellmann started the worldwide expansion of its international logistics network in 1982. It has opened additional offices, e.g., in Hong-Kong, China, Singapore, South Korea, Vietnam, Sri Lanka, and Japan. In 2004, it opened a large distribution center with its own terminal in the port of Shanghai, China.

At the same time, Hellmann expanded its logistics network in Europe and launched premium transport services with guaranteed deliveries within 24 and 48 hours. In order to increase the service level by more reliable and even faster deliveries, Hellmann has advanced the cooperation with other forwarding agencies. Therefore, since 1993 the *Night Star Express* alliance has provided overnight express services for documents, parcels, and packages in the whole of Germany. In order to guarantee fast and reliable transport of LTL freight all over the country, the *System Alliance* was founded in 2000 followed by the *System Alliance Europe* five years later.

Beside the classic business of forwarding agencies, Hellmann additionally founded the Personal Computer Organisation (PCO) in 1984 as a subsidiary company. Today, PCO provides logistics IT solutions such as track and tracing systems and automatic camera tracking software based on picture processing for goods in storage.

6.2.2 The Hellmann Office in Bremen

The Bremen office of Hellmann is located in the central freight terminal. It focuses on groupage traffic in North-West Germany. In order to increase reliability and decrease service times especially in the Western areas, a small-size office in Nortmoor was taken over from a subcontractor in 2009. This branch is managed from the office in Bremen. In the operating business, the Bremen office is responsible for postal codes 27 and 28. The whole catchment area is shown in Figure 6.2.

Beside general cargo logistics, the Bremen office provides services in the supply chain management and contract logistics for key and permanent customers. On the terminal premises there is also an office of the DPD which was a subsidiary company until 2001. To offer customers both general freight cargo and courier and express services, also small-size packages are collected in the pre-carriage of groupage traffic (cf. Section 6.1). These packages are unloaded at the depot and passed on to DPD which manages the further transport.

6.2.3 The System Alliance

In order to establish a reliable cooperation network which offers fast transport all over the country, Hellmann founded the *System Alliance* in 2000. The System Alliance provides special premium transport services with guaranteed delivery by 8am the next day. Its administration and management is located in Niederaula. In total, the *System*

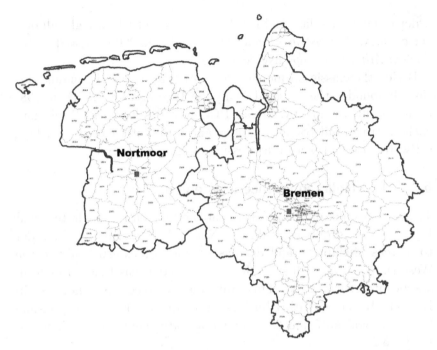

Figure 6.2: The catchment area of the Hellmann offices in Bremen and
Nortmoor.

Alliance includes 42 regional forwarding agencies and transport carriers with more than 10,000 employees and 6,000 vehicles which deliver
more than 9.4 million packages per year (System Alliance GmbH,
2014). The cooperation network is visualized in Figure 6.3. Participating companies have committed themselves to comply with multilateral
obligations and to respect the rules and instructions defined in their
partnership agreement. Therefore, the *System Alliance* ensures high
quality and reliability of transport. Due to these high standards,
quality-improvement measures significantly affect the internal processes of each single company. To a certain degree, the specifications
of the partnership agreement standardize the processes of the cooperating partners.

Figure 6.3: Members of the *System Alliance* in 2014 (System Alliance
GmbH, 2014, taken from p. 12).

The whole catchment area is divided into several subareas. Each subarea is assigned to a forwarding agency which is responsible for meeting the service standards and handling all transports within this area on time. Packages are either exchanged directly or by a single hub and spoke network (Cardeneo, 2008, p. 784) via a central hub located in Niederaula.

Key points which are specified and controlled by the System Alliance relate to:

- offered services and products,

- applied IT infrastructure, systems, and data structures,

- communication between forwarding agencies, partners, and customers,

- applied technologies for identifying packages as well as track and tracing systems,

- documentation and quality assurance.

The partners are committed to transport all packages and items which fulfill a specified length, height, volume, and weight. In general, the order situation is heterogeneous and not limited to pallets. Conventional orders are delivered in between 24 and 48 hours. In addition, there are so-called premium services which have to be delivered by 8am, 10am, 12am, or later the next day, while *flex* services will be delivered within 72 hours. High penalties for exceeding time windows in premium services motivate the participants to fulfill their commitments on time.

6.3 Process Documentation and Analysis

For three months all relevant processes in groupage traffic were documented and analyzed in the Bremen office of Hellmann Worldwide Logistics. Therefore, Max Gath passed though the company's different departments, namely warehouse, service, dispatching, business management, controlling, customs clearance, and shipment completion. Since the transport processes are outsourced, he also participated in pickup and delivery tours in pre- and onward carriage of subcontracted freight carriers. In all departments, several employees as well as the department mangers themselves supported him in his investigations.[1]

6.3.1 Business Process Modeling

In order to model the business processes, the *Architecture of Integrated Information Systems* (ARIS) approach was applied (Scheer, 1999). Therefore, the case study started with a detailed documentation and analysis of relevant processes. The well-established *Business Process Modeling Notation* (BPMN) was applied to reveal the interdependencies between processes, IT-systems, and actors. BPMN is an international standard for process modeling which has been developed by the *Object Management Group* (OMG) in cooperation with IBM, Oracle, SAP, and numerous other companies. It allows to clearly assign activities and tasks to actors and to transparently model the flow of information. Thus, this section determines all relevant actors and decision makers. It investigates how their decisions affect the processes and which data and information provide the basis for their decisions. A subset of elements and notations of BPMN which is used for business process modeling in this chapter is shown in Figure 6.4.

In order to cover the general planning and control in groupage traffic, the collected information has been enriched by interviews with other transport service providers and logistics experts. While the dispatching processes of different forwarding agencies vary in detail, e.g., by applying different software systems, the general procedure does not differ substantially. The main processes of pre- and onward carriage in groupage traffic are illustrated in Figure 6.5.

---

[1]Special thanks to Hellmann Worldwide Logistics & Co. KG and particularly to Jens Engelmann, Sven Bünger, Reinhard Riefflin, and Tobias Kohsman for their great cooperation and support.

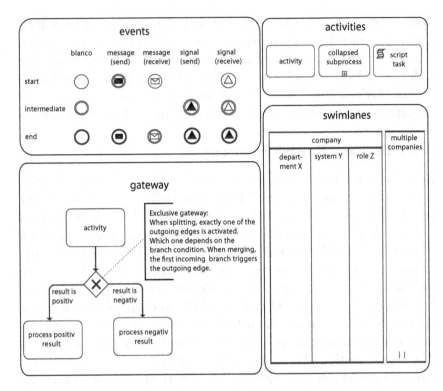

Figure 6.4: Elements and notations of BPMN which are used for business process modeling in this chapter.

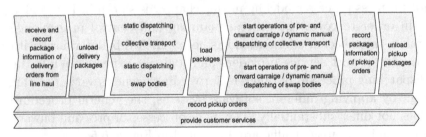

Figure 6.5: The main processes in pre- and onward carriage.

Similar to dispatching full-truckload freight, the planning and control of swap bodies especially in short-haul follows recurring patterns, shows less dynamics, and is less complex than dispatching the collective transport of multiple small and medium-size goods. In general, only regular customers require swap bodies which can be picked up and delivered in fixed tours on a daily or weekly basis. In addition, the number of possible and reasonable stop combinations is comparatively small and can be handled manually by a human dispatcher with near-optimal results.

Therefore, this section focuses on the dispatching process for collective transport which is illustrated in Figure 6.6. Firstly, an information system collects all incoming orders to assign them to predefined tours by a static mapping, e.g., of postal codes to tours. In general, the first assignment neither considers the amount of effectively received orders, nor the properties of packages, nor available vehicles, but is an essential pre-processing step. Secondly, this allocation is optimized by the dispatcher. He ensures that time-critical orders are processed with higher priority, identifies orders on overloaded tours, and reassigns these orders to tours that still have fee capacities. After this rough planning each contracted transport service provider starts a fine-grained planning as shown in Figure 6.7. In this step, the transport service provider schedules its trucks and assigns the orders to its vehicles.

The freight carrier determines the shortest route by applying his expertise and additional knowledge, e.g., about preferred time slots at the customers' incoming goods departments. In general, this is performed manually. Next, the freight carrier reports damaged or time critical orders which cannot be processed within guaranteed time windows to the dispatcher of the forwarding agency. Thus, the forwarding agency might instruct external transport providers to transport these time critical goods.

Figure 6.7 depicts the fine-grained planning of the transport service provider in detail. Firstly, the dispatcher checks if all of his vehicles are indeed available or if some freight carriers fall out. In addition, he checks if all assigned packages arrived at the storage in time.

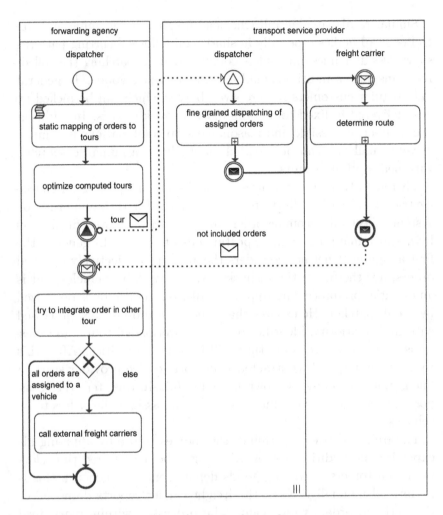

Figure 6.6: The planning process of a forwarding agency.

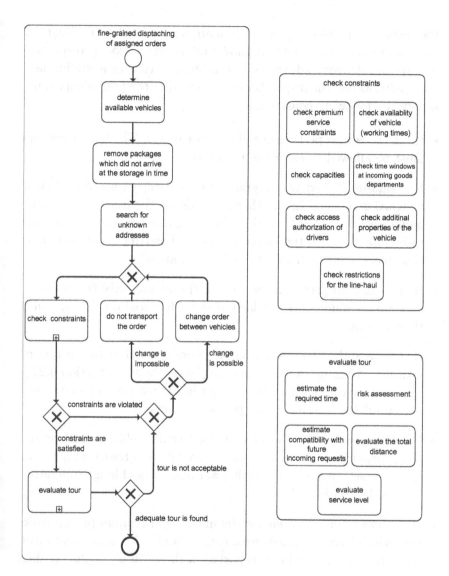

Figure 6.7: The fine-grained planning of transport service providers.

Packages which are delayed in the main leg are neglected. Next, he determines customers with unknown addresses, looks up respective pickup and delivery addresses in a map, and creates a preliminary tour. Afterwards, the dispatcher checks whether the preliminary tours satisfy the following constraints.

- He checks access permissions of freight carriers. The freight carriers must be allowed to enter the premises.

- He validates if used vehicles have sufficient capacities to load their assigned goods and fulfill other restrictions. For instance, the maximum weight and length of the vehicle must be below a certain value to be able to reach the incoming goods departments of particular customers (e.g., in inner city pedestrian areas).

- He checks if all premium service constraints are satisfied. Some orders are premium services which must be delivered within particular time windows.

- He verifies if the time restrictions at incoming goods departments are satisfied. Although some orders are *flex services* (cf. Section 6.2.3), some customers can only be serviced in certain time windows, e.g., only until 10am in inner city pedestrian areas.

- He checks the availability of vehicles. Certain vehicles must return to the depot in time, because they are required to transport goods in the main leg. In addition, varying working times of freight operators have to be considered.

- He checks the restrictions for the main leg. Premium pickup packages which have to be delivered, e.g., to South Germany, must be at the depot in time, to be transported in the main leg in the evening.

If a tour violates a constraint, the dispatcher tries to change orders between tours. Otherwise he evaluates each tour by the following criteria.

- The dispatcher estimates the time which is required to service all orders.

- He performs a risk analysis, e.g., by considering possible delays at incoming good departments or through traffic congestion.

- He evaluates the efficiency of tours by considering the total kilometers driven.

- He determines the service level. For instance, it is preferable to send the same driver to regular customers, because he will then know customer specific demands and further restrictions, e.g., at incoming goods departments.

- He estimates the flexibility of the tour. The goal is to determine a tour in which later on incoming pickup orders can be integrated dynamically.

If possible, the dispatcher further improves tours. Unserved orders are returned to the dispatcher of the forwarding agency.

In conclusion, the general planning and control processes in groupage traffic reduce the problem complexity by splitting up the overall problem (the assignment of all orders to vehicles while considering relevant constraints) into smaller, less complex problems. After the pre-processing, each contracted transport provider solves the reduced problem of assigning a subset of orders to a subset of vehicles. However, this implies that possible dependencies and optimization potentials between orders of different transport companies are not detected and consequently neglected.

Next, the data which serves as the basis for the decisions in autonomous processes is identified and their format, amount, quality, feasibility, relevance, and point of operation where the data is available are analyzed. Processing only data which are available in real-world processes allows for the integration of autonomous logistics processes without any investment in new hardware. For instance, software systems should not consider the exact volume as there is no reliable information available in current information systems, because so far the volume of heterogeneous goods cannot be determined automatically.

In order to keep confidential data of Hellmann private and to prepare a more generally valid documentation of the planning and control processes in groupage traffic, this thesis conceals detailed data structures but considers abstract information which matches the processes of freight forwarding agencies in general. Figure 6.8 provides a legend of the applied notations and elements used in the Extended Entity Relationship (ER) diagrams presented in this section. In ER diagrams, information and data are consolidated in entities (perceptible or imaginable objects). Data attributes are assigned to these entities. In this diagram, optional attributes refer to data with restricted availability. Therefore, tour planning algorithms should neglect this information. Figure 6.9 depicts an ER-diagram including the data which form the basis for the dispatching processes in pre- and onward carriage.

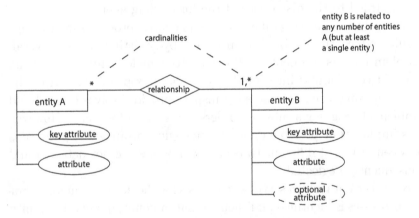

Figure 6.8: Elements and notations of Extended Entity Relationship diagrams which are presented in this chapter.

The most relevant data is the *package data*, because this information is precise knowledge of the packages which arrive from the main leg.

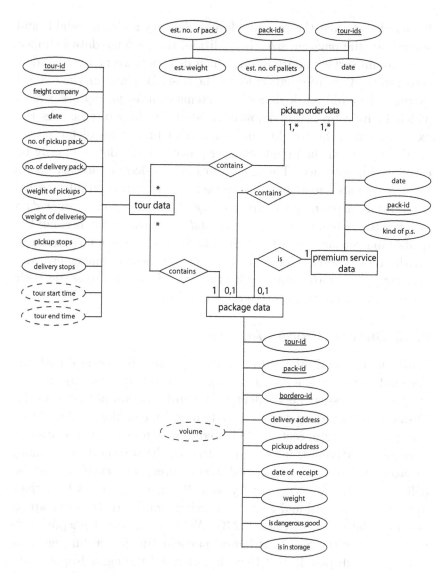

Figure 6.9: The Extended Entity Relationship diagram of relevant data and information which can be processed for dispatching in groupage traffic.

It includes at least the pickup address, delivery address, weight, and sometimes the package size. In addition, the package data is linked to premium service information, which defines delivery time window constraints. This information forms the basis for planning the onward carriage. In addition, logistics IT systems contain *pickup order data*, which include uncertain knowledge about pickup orders. As the exact amount, the weight, and additional properties of picked up goods depend on the manufacturing process, these data can only be estimated in advance. The exact data is revealed at the customer's incoming goods department. Nevertheless, the estimated information is essential for panning the pre-carriage. Later, the *pickup order data* is linked to the *tour data* and *package data*. Not the dispatcher, but the quality management uses the *tour data* for performance measurement. Furthermore, departments such as the business management and accountants' department require the *tour data*, e.g., to settle their accounts with the transport service providers.

6.3.2 Business Process Analysis

With the data and information shown in Figure 6.9, several perform-ance indicators were aggregated to quantitatively describe the current stages of processes and to identify the optimization potential in the business processes related to dispatching. One of the results of this analysis is that total costs could be reduced by minimizing the amount of required external transport providers for the transport of premium services packages. As described above, premium services must be delivered on the following day by 8am, 10am, 12am, or not later than 5pm while conventional and flex services can be distributed up to two days later (cf. Section 6.2.3). Within logistics transport net-works (such as the *System Alliance*) participating forwarding agencies must pay high penalties if they do not fulfill the agreed-upon com-mitments. Therefore, forwarding agencies frequently order external transport providers to transport premium service packages which have to be picked up or delivered urgently. In the Bremen office of Hellmann, external freight carriers transported about 6% of all

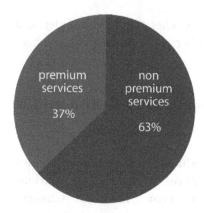

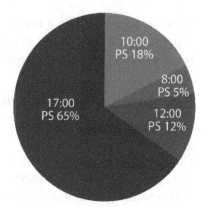

Figure 6.10: Proportion of premium services and non premium services delivered or picked up by external transport providers.

Figure 6.11: Types and percentage of premium services delivered by external transport providers.

orders in 2011. Moreover, the orders serviced by external transport providers were analyzed. Figure 6.10 shows the proportion of premium services and non premium services in 2011. Figure 6.11 sub-classifies these premium services and denotes their respective percentage of all premium services transported by external transport providers.

The figures reveal a high savings potential by awarding higher priority to premium services than to conventional orders in the dispatching process. Conventional orders and flex services can be delivered up to two days later and still be within guaranteed service time. By shifting the transport of these packages to the next day, own vehicles can process packages with higher priority. Thus, the amount of external transport providers might be reduced.

However, the consideration of all preferences and constraints is complex. The complexity is even increased by the high degree of dynamics. Tours and routes have to be adapted throughout the entire operation, to include new orders in existing plans and to deal with unexpected events. As a result, the dispatcher requires solid decision support by an information system that is capable of fulfilling the high requirements in groupage traffic.

6.4 Simulation of Multiagent-Based Autonomous Groupage Traffic

Changing logistics processes and introducing new technologies such as multiagent systems (MAS) often requires investment, negotiations, and intensive communication with the persons involved. Changes affect not only work sequences of in-house departments, but also operational procedures of contracted companies, e.g., external transport providers. Multiagent-based simulation (MABS) reduces the company's risks and procures well-founded assessments of the impact of potential changes (Pokahr, Braubach, Sudeikat, Renz, and Lamersdorf, 2008, p. 11). Thus, it prevents investments if simulated scenarios reveal a lower benefit than expected. Furthermore, a fine-grained simulation allows for detailed previews and predictions about the effects of alternative planning strategies and processes. Therefore, MABS identifies and evaluates characteristics of a new planning and control system before its implementation in the company. This is especially relevant in scenarios where the quality of the result depends on the outcome and/or sequence of agent negotiations that cannot be predicted in advance (Jennings, 2001). Moreover, it facilitates for precise strategical analysis and the investigation of exceptional situations. For instance, effects of new pricing models or the impact of economic cycles and natural disasters on the supply chain can be determined.

As a result, MABS is applied to substantiate that the dispAgent approach considers customized demands and optimizes dynamic and complex real-world planning and control processes in groupage traffic by the implementation of multiagent-based autonomous processes.

6.4.1 The PlaSMA Simulation Framework

There are multiple platforms and tools for MABS, e.g., JadeX[2] (Pokahr, Braubach, Walczak, and Lamersdorf, 2007; Pokahr and Braubach, 2009) which facilitates implementing BDI-agents (Braubach, Pokahr, and Lamersdorf, 2005) and enables simulating workflows and processes described by BPMN (Braubach, Pokahr, and Lamersdorf, 2013a). Another professional tool which supports simulation and execution of large scale MASs is Aimpulse Spectrum[3]. While these are general platforms and not specialized for any domain, this section focuses on the PlaSMA Simulation Framework (Warden et al., 2010) which has especially been developed for, but not limited to, simulating dynamic scenarios in transport logistics.

The *Platform for Simulation of Multiple Agents* (PlaSMA) is an agent-based event-driven simulation platform which was developed within the *Collaborative Research Center 637 – Autonomous Cooperating Logistic Processes* (SFB 637) by the Institute for Information and Communication Technologies (TZI) at the University of Bremen.[4] It is designed for modeling, simulation, evaluation, and optimization of dynamic planning and control processes in logistics. It extends the FIPA-compliant Java Agent DEvelopment Framework (JADE) (Bellifemine et al., 2007) for agent communication and coordination. PlaSMA provides discrete time simulations, which allow precise simulations of processes with small simulated time intervals (with intervals of at least 1ms). Furthermore, it ensures correct synchronization and reproducibility (Gehrke et al., 2008). For instance, the simulation framework guarantees that message transfer consumes simulated time, because transferring messages consumes physical time in real-world processes as well. Consequently, the consistency of each agent (e.g., no agent receives messages from the future and all the agents' knowledge is consistent at a certain point of simulated time) is also guaranteed

---

[2]For more information `http://www.activecomponents.org` (cited: 1.9.15).

[3]For more information `http://www.aimpulse.com` (cited: 1.9.15).

[4]A professional advertising film of the simulation platform is provided at `http://plasma.informatik.uni-bremen.de` (cited: 1.9.15).

by conservative synchronization mechanisms (cf. Gehrke et al., 2008, for more details). The time model adequacy is ensured by a parameter controlling the maximum and minimum simulated time interval for the synchronization. Thus, PlaSMA is capable of simulating scenarios which require both fine-grained and coarse time discretization.

In order to reliably evaluate logistics scenarios and to perform pertinent strategic analysis, PlaSMA was improved for this case study to model real-world infrastructures and to support their import from OpenStreetMap[5] (OSM). Firstly, the implemented import tool loads OSM data. Secondly, it reduces the original number of nodes by removing nodes which connect exactly two edges with the same properties. These nodes only exist to visualize the course of the path but can be neglected for creating a precise model of the infrastructure. Thus, the two connected edges are consolidated in a single one. The removal of irrelevant nodes and edges significantly reduces the memory requirements to save complex real-world infrastructures. In order to prevent deadlocks caused by inaccurate data, nodes that cannot be reached from or to the depot of the transport service provider are removed. Therefore, an efficient shortest-path search based on the algorithm of Greulich et al. (2013) was implemented for this case-study. The algorithm performs a complete Dijkstra search (Dijkstra, 1959) on the infrastructure graph by applying radix-heaps (Ahuja, Mehlhorn, Orlin, and Tarjan, 1990) and determines unconnected nodes. Beside roads, the tool allows the import of waterways as well as railway networks. Figure 6.12 depicts the graphical user interface (GUI) of the OSM importer tool. As a result, PlaSMA is capable of modeling large scale arbitrary real-world infrastructures with more than 300,000 ways, e.g., roads, motorways, trails, and waterways, and 200,000 traffic junctions on a laptop computer. Beside its type, e.g., inner city road, country road, motorway, or walkway, each way includes additional information about its length, one way restrictions, and speed limit. Since even road sections with varying speed limits are differentiated, PlaSMA supports the simulation of fine-grained and

---

[5]For more information see http://www.openstreetmap.org (cited: 1.9.15).

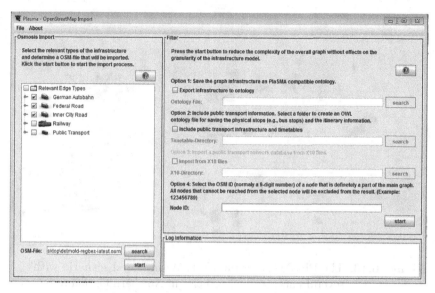

Figure 6.12: The graphical user interface of the OSM importer which allows the import of real-world infrastructures for accurate and precise simulations by memory-efficient graph representations.

pertinent logistics scenarios. Other projects also benefit from this new function such as the projects of Gath, Wagner, and Herzog (2012), Greulich, Edelkamp, Gath, Warden, Humann, Herzog, and Sitharam (2013), Greulich, Edelkamp, and Gath (2013), and Edelkamp, Gath, Greulich, Humann, and Warden (2014).

Moreover, PlaSMA was extended to link process and customer data of industrial partners directly with the simulation platform. It is essential to import real service requests and real available transport providers of simulated days automatically and at run time, to induce plausible and precise results that permit conclusions and analysis of real logistics processes. Thus, it is mandatory to resolve real addresses to corresponding nodes within the modeled transport infrastructure. Consequently, a Geographic Information System (GIS) is implemented which is based on OpenStreetMap and Google Maps services. The

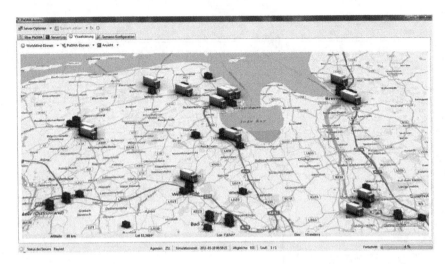

Figure 6.13: The PlaSMA user interface. A simulated scenario with several packages and vehicles in Northern Germany.

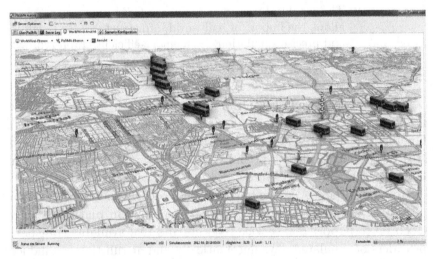

Figure 6.14: The fine-grained infrastructure model of the city of Bangalore, India.

system starts a reverse geocoding process to determine the coordinates of addresses and applies a nearest neighbor search to identify the nearest node within the modeled infrastructure.

PlaSMA was further extended to support batch-runs. Also, a ratio system for automated measurements of individually defined performance indicators was integrated. This allows to optimize the dispAgent software system's performance and to investigate its behavior by comparing performance indicators in numerous simulation runs.

For the demonstration of autonomous logistic processes, the graphical user interface (GUI) is also enhanced, e.g, by an adequate infrastructure representation which depends on the zoom level (cf. Figures 6.13–6.14). The visualization improves traceability and allows the investigation of dynamic interdependencies within the process over time.

6.4.2 Simulation and Experimental Setup

In order to allow fine-grained simulations which reflect real-world conditions, the modeled road network covers the whole trading area of the Bremen office of Hellmann in Northern Germany. The modeled transport infrastructure contains 156,722 traffic junctions and 365,609 roads. It includes all relevant highways, motorways, and inner city roads of the OpenStreetMap database. To verify the system's performance quantitatively with comparison to the real-world processes and to show its applicability, the simulated scenario includes the effectively transported orders (about 1,100 per day on average) as well as the real fleet operated within the simulated time window of the representative time period from 1 to 10 November 2011.

As mentioned above, the investigation neglects full-truckload freight and orders transported by swap bodies, because the planning and control of swap bodies follows fixed recurring patterns. It is less complex and less dynamic than the dispatching of collective transport. Instead, all orders which were dispatched for collective transport as well as all orders which were delivered by external freight carriers have been considered.[6] Each order is given the unique characteristics the real shipment it represents such as pickup and delivery location, weight, time windows, as well as premium service constraints. Moreover, it has information of the incoming date on which an order was available for dispatch. Since exact incoming dates with timestamps of pickup orders are not available, only the date is considered. Thus, the dynamic handling of pickup orders is modeled by setting the incoming date of every 10th order to a random time of the day during operation. The modeled fleet reflects real speed limits as well as capacities. Note that the simulated velocity of the vehicle is reduced by the speed limit on the road sector the vehicle is driving on. In addition, the maximum speed limit on the road sector is reduced by 20% to cover real-world average velocities on the roads in the simulation. For instance, if the vehicle has a speed limit of 130 km/h and the speed limit of the road section is 80 km/h, the vehicle has an average velocity of 72 km/h on this road section in the simulation. Furthermore, it is assumed that the vehicle leaves the depot at 7am and has to be back at the depot at 5pm at the latest. Vehicles which return to the depot later than 2pm remain at the depot and do not start an additional tour. Vehicles have a speed limit of 80 km/h. In addition, we simulate that each cargo handling operation of shipments (up to 300 kg) takes 15 minutes.[7] All assumptions were deduced from real-world processes and were checked by experts at Hellmann to ensure that they reflect real-world conditions.

The dispAgent software system is configured to consider the domain-dependent requirements of groupage traffic. Firstly, the pre-processing is adapted. The assignment of vehicles to clusters considers a company specific mapping of vehicles to tours and postal codes. Consequently,

---

[6]About 0.3% of all orders found in the data pool cannot be considered in the simulation because the raw data is too incomplete and inaccurate for simulation. Probably these orders are *dummy* entities, because they are not linked to other data sets providing detailed order information. In addition, pickup orders with pickup addresses and delivery orders with delivery addresses which are beyond the service area (and mostly out of Germany) are neglected.

[7]With the exception of handling operations at the depot.

each freight carrier has a fixed service area. In real-world processes this is highly relevant, because the driver has special knowledge about customized processes and sometimes personalized access authorizations at the incoming goods departments of the customer. Furthermore, it increases the service quality and enhances the customer's loyalty.

Secondly, the decision-making process of the vehicle agents is configured to minimize the total number of required external transport providers. The system classifies orders either as a conventional order or as a premium service order. A premium service order must be delivered on the day of operation. Conventional orders can also be delivered the next day or even later. Since pickup orders have to be picked up on the same day, the system automatically classifies them as premium services. Consequently, the agents consider these premium service constraints as follows (cf. Section 4.4). The vehicle agents' highest priority is to maximize the amount of transported premium services. Its second highest priority is to maximizes the number of transported conventional orders and the third highest is to minimize the total distances driven. Thus, a vehicle agent might prefer to transport a single premium service instead of multiple conventional orders if the capacities are exceeded or the remaining time is insufficient to service all orders. As either pickup or delivery point of each order is the depot, all depot stops are consolidated in a single stop. As a result, the size of the problem is comparatively small. Problems generally contain less than 15 stops. Consequently, only the branch-and-bound algorithm is applied in the agents' decision-making processes. In general, the simulated real-world problems are highly constrained and the pruning rules reduce the search space significantly. Therefore, the algorithm is configured to terminate after 300,000 expansions at the latest in order to increase the number of multiagent-negotiations.

Agents representing unserved orders start new negotiations with vehicle agents if the order has not been loaded yet. Continuous negotiations between vehicles to further optimize an allocation if a feasible solution is found are not implemented, because this function was not included in the version of the software used for the case study. Nevertheless, it would further increase the quality of solutions.

6.4.3 Results

The dispAgent approach supports dispatchers as well as contracted freight carriers during operations. It automatically controls the processes and adapts plans if a delay or a changing order situation is detected. The system is designed to meet the special requirements in groupage traffic. It supports the combination of pickup and delivery tours without exceeding the maximum capacity of vehicles and considers all relevant constraints such as time windows, handling times, as well as request priorities. Moreover, the system maximizes the number of transported premium services as well as the processed amount of conventional orders. It minimizes the length of routes for each vehicle.

Since in current processes each freight carrier determines its route manually, the system increases the efficiency by providing optimal and factual proposals at the start of each shift. It checks hard constraints automatically and accelerates the decision-making of freight carriers. As a result, each freight carrier saves about 20 minutes of time each day. Moreover, the continuous process monitoring improves transparency. The current positions of each shipment can be visualized and additional information about each load is provided, e.g., actual weight, estimated time of arrival, and volume. This information can also be applied to further optimize and synchronize the supply chain. For instance, the estimated time of arrival can automatically be sent to the incoming goods departments of customers (e.g., via apps), who start preparing the receipt of goods or proactively send a message indicating that a delay is expected. Optionally, this information can directly be sent to *smart* production facilities and factories which are connected to the internet and can react autonomously to disturbances, delays, and changes. In addition, freight carriers can also react in advance to the changing situation and adapt tours and routes if necessary. The dispatching system increases the customer service level by reliable pickups and deliveries. At each step of the process the system checks the time window and the premium service constraints.

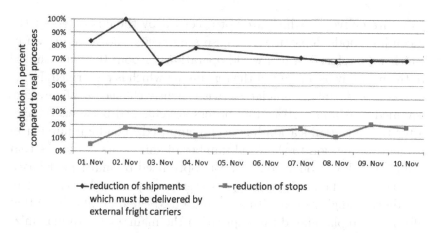

Figure 6.15: The reduction of stops and shipments which must be transported by external freight carriers on each day of the simulated time period compared to real-world processes.

The results of the simulated scenario reveal an increasing efficiency and a significant reduction of costs by applying the multiagent-based dispatching system in groupage traffic. For instance, the number of stops is reduced by an average of about 15%. The multiagent system enables the efficient grouping of packages at a certain pickup or delivery location. Consequently, loads at the same location are transported by a single vehicle (if possible). If a vehicle has a delivery stop, the dispAgent software system ensures that a potential pickup at the same customer location is done by the same vehicle, even if this could result in replanning and in shifting a conventional order to the next day.

Moreover, the number of shipments which have to be transported by an external transport provider is reduced by an average of 75.64%. This is due to consequently prioritizing the transport of premium services in the planning process and to shifting the delivery of conventional orders to next days if a premium service or pickup order can be delivered instead. On average, about 10% of all orders are shifted to the next day. Approximately 3% (which are also included in the

10%) are deliveries which are not accepted by the customers at the incoming goods department. These are exactly the same deliveries which were refused in real-world operations. About 0.7% are shifted to the next day as a result of the replanning which is performed during operations. A similar simulation (with slightly changed configuration) revealed that about 50% of all orders are changed between vehicles after the initial allocation has been computed. The high number of changes in relation to the total number of shipments, indicates that the initial allocation is continuously optimized by multiagent-based negotiations. Thus, the multiagent system improves the reaction to the daily changing order situation and unexpected events. Note that also in the replanning during operation the main goal is to maximize the amount of premium services.

Figure 6.15 depicts the reduction of stops and of shipments which have to be delivered by external freight carriers on each day of the simulated time period in comparison to real-world processes. As a result, the dispAgent software system saves costs of several thousand Euros per day by increasing the capacity utilization of the company's own vehicle fleet and thus reducing the required number of external transport providers.

Finally, all tours and routes of an arbitrary day were investigated in detail in cooperation with the dispatcher at the Bremen office. Therefore, a map of all tours and routes in Google Earth was generated which includes all stops and additional information such as time of arrival, weight of the freight, number of pickup or delivery units at a stop, the maximum capacity of the truck, the working time of the freight carriers, total delivery weight of a tour, total pickup weight of a tour, etc. An example is provided in Figure 6.16. In order to clearly show and analyze interdependencies between multiple tours of the day, Appendix B.1 visualizes all tours of an arbitrary day on a single map. In addition, Figure 6.17 allows analyzing the kilometers driven by the vehicles.

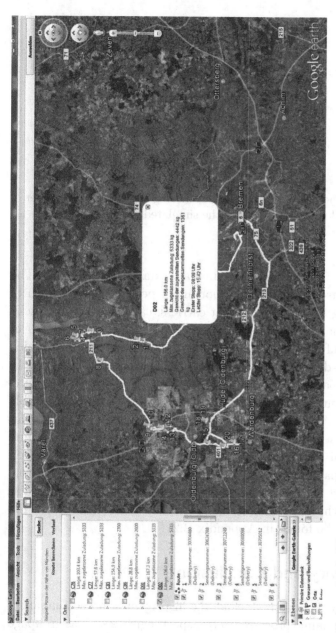

Figure 6.16: The visualization of an arbitrary tour in Google Earth.

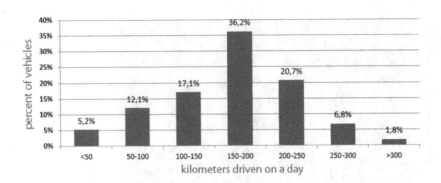

Figure 6.17: The percentage of vehicles driving the respective number of
kilometers per day in the simulated scenario from the 1 to
10 November 2011.

The basis of the analysis is a simulated day with 56 tours. The
discussion focused on several important aspects such as length of
routes, working times of the freight carriers, capacities of the trucks,
guaranteed time windows of orders, vehicle restrictions with respect
to transport infrastructure, as well as special situations in which
additional non-digitized knowledge of the dispatcher is essential. In
conclusion, the analysis checked the suitability for daily use of automatically computed tours and routes. Tours and routes are suitable
and realizable. The multiagent-based dispatching system automatically considers more constraints and options during the planning
processes. Thus, it increases the efficiency of allocation. However,
the human dispatcher is able to further improve proposals selectively
with his expert knowledge which is not represented electronically.
The protocol which documents the results of the discussion with the
dispatcher can be found in Appendix B.2.

6.5 Summary and Conclusion

This case study demonstrates that the dispAgent approach enhances
complex and dynamic processes in transport logistics such as groupage

traffic with an adaptive and reactive system behavior to optimize the efficiency and to increase the service level through reliable transport. It started with a general overview of the processes in groupage traffic. Next, the most relevant processes of the Bremen office of Hellmann Worldwide Logistics were investigated in detail. As the processes in groupage traffic are standardized by the *System Allicance*, the processes of other offices and companies participating in the cooperation network are similar to those of the Bremen office of Hellmann. On the basis of the process documentation and analysis, the planning and control of collective transport were identified as the most complex and dynamic processes with the highest potential for optimization. For instance, the consequent consideration of premium service constraints in the planning and continuous replanning of tours result in significant cost savings. Next, the dispAgent approach was configured to meet the special requirements in groupage traffic. A real-world scenario was modeled within the PlaSMA optimization framework. To this effect, PlaSMA was extended by a fine-grained infrastructure model to support the import of real-world road networks, by a performance measurement system, and by further components which facilitate the pertinent and accurate simulation of logistics transport processes. The modeled and simulated real-world scenario of the Bremen office of Hellmann proves that the developed dispatching system supplies freight carriers and dispatchers with suitable tour and routing proposals which fulfill the individual requirements of a company in groupage traffic. The system significantly decreases the daily costs by reducing the required number of external transport providers. It increases the efficiency and meets the special challenges in groupage traffic. Nevertheless, the overall efficiency can be further increased by the implementation of continuous trading approaches (cf. Section 4.3) which allow saving costs by changing orders between agents representing vehicles. They were not included in the dispAgent software system at the time the case study was performed. In addition, the human dispatcher can further improve proposals selectively with his individual expert knowledge which cannot be represented electronically.

7 A Case Study: Courier, Express, and Parcel Services

Beside groupage traffic, another challenging domain for logistics planning and control is urban courier, express, and parcel services (CEP services). This case study was performed in cooperation with tiramizoo GmbH in its headquarters in Munich, Germany. tiramizoo offers immediate same-day and reliable CEP services. In multiple simulated scenarios as well as in real-world applications, the case study shows that the software implementation of the dispAgent approach is able to handle the high complexity and dynamics of this domain. It optimizes the efficiency of the dispatching processes by reducing the number of required vehicles, the amount of stops, and the required time. Simultaneously, it improves the service quality by extending the product range and ensuring reliable deliveries. Both goals are achieved by handling the dynamics with an adaptive, reactive, real-time, and customized system behavior.

Firstly, Section 7.1 introduces CEP services in general. Secondly, Section 7.2 briefly presents tiramizoo GmbH.[1] Next, Section 7.3 analyzes the application of a standard software product for planning in the CEP services sector and presents the identified optimization potential. Section 7.4 provides a quantitative evaluation of the dispAgent approach in the CEP services domain, followed by a qualitative evaluation in Section 7.5. Finally, Section 7.6 summarizes the results of the case study.

---

[1]Special thanks to tiramizoo GmbH and particularly to Mateusz Juraszek, Jacek Becela, Michael Löhr, Thomas Bluth, Dirk Reiche, and Philipp Walz for their great cooperation and support.

7.1 Courier, Express, and Parcel Services

As shown in Chapter 1, globalization, the shift from seller to buyer markets, Industry 4.0, the strong growth spurt of e-commerce, and the resulting goods structure effect have changed the economic and logistics structures. As a result, there is a much higher amount of small-size shipments, which have to be delivered within guaranteed time windows and sometimes within a few hours. In addition, the demand for customized logistics services increases. Especially, CEP services profit from this trend. The turnover of this industrial sector has more than tripled over the last ten years in Germany (Cardeneo, 2008, p. 782). The parcel volume of CEP services was more than 2.7 billion in 2013 (Esser and Kurte, 2014, p. 6). It is expected that the volume will have increased to over 3.2 billion packages per year by 2018 (Esser and Kurte, 2014, p. 13). Moreover, within Industry 4.0 concepts, CEP services will be of particular importance as part of last mile logistics between production and end-users. In Western Europe, the amount of parcels will increase by 4% per year in the B2C market until 2020 while the amount of same-day deliveries will increase to over 3.2 billion packages per year (Netzer, Krause, Hausmann, and Hermann, 2014, p. 5 and p. 16).

However, the rising demand for reliability and immediate delivery increases the complexity and dynamics of planning and control processes of CEP services providers (Fleischmann, 2008, p. 8). In groupage traffic or container logistics, the freight is transported via central depots or hub and spoke networks (Cardeneo, 2008, p. 785) as presented in Chapter 6. In contrast, urban CEP services providers transport goods from the sender to the receiver via multi-stop networks (Vahrenkamp, 2007, pp. 143) in direct tours. There is no central depot. Furthermore, the characteristics of the transported freight are different. In groupage traffic and container logistics, the number of orders is low and the load weight is high. CEP services providers, however, transport larger amounts of goods with lower weight (Vahrenkamp, 2007, p. 137). Due to the highly volatile order situation and the demand for immediate transport, the number of used vehicles is also varying during daytime operation. Therefore, new vehicles are added and others leave during the day.

In general, the diversity of vehicles ranges from bicycles, e-bikes, cargo bikes, e-cargo bikes, scooters, motorcycles, cars, wagons, and vans from 2.8 tons to trucks with a total weight of 7.5 tons. In urban districts, the use of bicycle couriers, a particularly sustainable form of courier and express services, is increasing (Manner-Romberg, Symanczyk, Ströh, Deecke, Bastron, and Marwig., 2009, p. 55). Bicycle couriers reduce CO_2 emissions as well as traffic volume. Especially in congested overcrowded areas they can reduce the service time. E-bikes and e-cargo bikes can also be used to transport heavy goods over large distances. For example, with today's e-cargo bikes it is possible to transport a weight of 120 kg at a speed of 25 km/h over a distance of up to 130 kilometers. The use of e-bikes reduces pollution, traffic volume as well as greenhouse gases in urban districts and mega cites. However, their capacity and range are limited. For the transport of large and bulky goods, the use of fully motorized vehicles remains essential.

Another difference to general cargo transport is the organizational structure of the vehicle fleet. In groupage traffic and container transport, the forwarding agency itself or subcontractors guide the vehicles and freight carriers. Their remuneration is mostly fixed. In contrast, the pricing systems of CEP services are generally commission-based. Thus, the autonomy and self-determination of freight carriers must be considered in negotiations and in the planning process. Also freight carriers within a company may compete with each other for profitable orders. Nevertheless, cooperation is required to deliver all goods reliably and on time. As a result, it is essential to consider the customer demands as well as individual properties and requirements of the heterogeneous and autonomous vehicle fleet, which is paid on a commission basis. For instance, in highly frequented business and industrial districts tours can be profitably consolidated while in *out-of-the-way* areas this is not possible. Thus, in contrast to the collaborative tour planning in groupage traffic and container logistics (cf. Chapter 6), the competition between couriers has to be considered by the dispatcher and applied decision-support systems. Nevertheless, cooperation and consolidation remain essential to reduce overall costs and to increase efficiency. The dispatching is made more difficult by the short panning horizon and high degree of dynamics. Most orders arrive during day-time operations. Even if multiple goods have to be picked up from or delivered to regular customers, interviews with several experts of the sector revealed that only about 20% of all orders are known one day in advance. In addition, exact time windows of regular orders are also unknown. Both factors increase the complexity of the planning process.

Similar to groupage traffic and container logistics, IT-systems are implemented to support the order management and to localize the couriers. There are also professional IT-Systems which compute allocations of orders to couriers. However, depending on the particular problem the computed tours have to be rechecked, adapted, extended, and optimized (in most cases manually) to consider all relevant customer constraints and dynamic events. Sometimes business processes are even modified because of limited capabilities of the IT-systems applied. Due to the high complexity and dynamics, the dispatcher is not able to consider all constraints and objective functions for optimal resource utilization. This holds especially true for real-time allocation. For instance, to ensure the autonomy and self-determination of couriers, a mixture of hierarchical centralized job assignments based on formal rules and a decentralized self-determining order acceptance is implemented in real-world operation (Weddewer, 2007, pp. 35).

In conclusion, it is essential for the automation of planning and control processes of CEP services providers to consider customized constraints and to fulfill the increasing requirements of reliability and prompt delivery in these complex and dynamic domains. Efficient tour proposals relieve freight carriers as well as dispatchers, increase resource utilization, and minimize costs.

7.2 tiramizoo GmbH

The tiramizoo GmbH was founded in 2010 and is one of the leading providers of logistics software and CEP services for same-day delivery in Germany. It offers prompt, reliable, and fast deliveries in more than 20 major cities. On its online platform, tiramizoo offers local delivery with fixed time windows within a few hours to dealers in consumer electronics, groceries, spare parts, tools, and construction materials (to name but a few). Optionally, the dealers can implement the same-day delivery option directly in their business processes, e.g., in their local shops by mobile devices and apps or in the ordering process of their online platforms by a plug-in. As the delivery of goods which are ordered at established online platforms requires between one and three days, end customers as well as local companies benefit from prompt delivery which increases their competitiveness considerably. The core business focuses on fully automated order acceptance, which is customized to the dealers' business processes, as well as automated assignment of orders to urban couriers via apps.

Growing cost pressure and the demand for increased service quality require the computation of cost-efficient tours which allow low-cost deliveries within guaranteed time windows. Thus, fully automated planning and control is one of the most relevant processes in the company. In order to handle the high dynamics, tiramizoo divides the working day into time segments of two and three hours. Each of these segments is considered as an independent problem which is solved separately. A professional dispatching software computes the tours for the next segment. Subsequently, the freight carriers start their tours at the depot(s). This process is repeated for each time-segment. As a result, the dynamic continuous problem is split up into multiple episodic static problems. Dynamically incoming orders are shifted to next time segments. Local disturbances during operation are handled by the freight carriers themselves. However, the transformation of the continuous dynamic problem to multiple static episodic problems requires additional synchronization of the processes for the transition from the last segment to the next one. On the one hand, it has to be ensured that all vehicles are back at the depot on time, because they must simultaneously start their tours at the beginning of the next segment. On the other hand, returned vehicles have to wait until the next segment starts. Moreover, the service quality can be increased further because the episodic problem structure prevents customized delivery times which range over several time segments.

Since standard dispatching software does not consider customized requirements, pre- and post-processing procedures are implemented which transform the input and output data to be processed by the standard software. For instance, these procedures allow varying handling times at the customers' incoming goods departments and/or freight dependent handling times. Moreover, they ensure that security measures of some customers are satisfied which may require entering the customer's premises only if the vehicle is empty or if it is exclusively loaded with goods from this customer.

7.3 Optimization Potential

For planning and scheduling CEP services the standard dispatching software which was applied by tiramizoo has the following drawbacks.

- It does not support time-distance matrices, which define the travel times between pickup and delivery locations. In order to compute travel times from one stop to another, it only considers distances and a unified average velocity of a vehicle to compute the travel time. Thus, neither vehicle dependent road access nor exact (vehicle dependent) conditions of road network sections can be considered which both effect the real travel time. Since the average velocity in urban districts and on highways differs substantially, the impact on the solution is significant.

- The handling time for the loading and unloading of shipments has to be denoted in the order's properties and is considered with this duration by the planning algorithm. However, the handing time is nearly the same if an additional shipment has to be loaded or unloaded at the same stop. Only if the amount of handled goods grows significantly, the required time increases remarkably. Therefore, it is necessary to differentiate the estimated service time. The estimated time should consider the amount of shipments which have to be picked up or delivered at a certain location. In addition, the handling times might depend on the specific conditions at the customer's incoming goods department and on the nature of the goods.

- Due to the fact that vehicles are loaded when they start their tours at the depot, handling times at a depot are neglected by the tour planning algorithm. However, if the vehicle has to pick up goods at a depot which is not the starting point of the tour, it is essential to include the handling time in the planning.

- The professional software product does not ensure that a vehicle has to be back at the depot at a predefined (individual) time.

In order to satisfy the specific requirements of tiramizoo, the software system is customized with pre- and post-processing procedures. In the pre-processing, the distance matrix is scaled to ensure that driving on inner-city roads as well as on highways at the same average speed reflects real-world conditions. Therefore, the pre-processing artificially enlarges inner-city roads and shortens highways. Let *speed* denote the real-world average speed for driving the *distance* with an average velocity *avgVelocity*. The distance is scaled by

$$scaledDist = \frac{distance * avgVelocity}{speed}. \tag{7.1}$$

However, as the distance matrix is the same for all vehicles, vehicle dependent travel times cannot be precisely considered. Furthermore, the pre-processing shortens the delivery time windows of orders to ensure that there is sufficient time to return to the depot. In addition, the handling times of orders with the same pickup or delivery location are shortened to avoid a single stop requiring an extraordinarily long time. However, if these orders are transported by different vehicles, because their pickup or delivery stops are located in the opposite direction, the handling times are underestimated. Therefore, the post-processing rechecks and adjusts the computed tours of the standard software product. Unified travel times between stops as well as handling times are substituted by more precise times. As a result, travel and handling times might increase through the post-processing. Thus, the post-processing might violate constraints and cause delays at customer locations as well as at the depot when the vehicle returns.

Due to the fact that the above mentioned constraints are not directly considered in planning, there remains an optimization potential which can be exploited by the dispAgent approach. As shown in Chapter 4, a vehicle agent can already consider all the above mentioned individual properties and requirements in its decision making process by tour planning algorithms. Thus, no pre- and post-processing of tours is required, constraints are satisfied, and the rising optimization potential is exploited in the tour construction process.

Moreover, the transformation of the continuous problem into multiple episodic problems is not necessary if the software system handles the dynamics. Thus, synchronization between episodic problems is not required, which offers further potential for optimization. In addition, the increased flexibility of delivery time windows which span over several time segments increases the customer service level.

7.4 Quantitative Evaluation

This section compares the system performance of the standard software product, which applies multiple algorithms (i.a., a large neighborhood search), including the required pre- and post-processing, with the performance of the dispAgent software system. The systems solve real-world problem instances from multiple service areas of tiramizoo with up to 50 orders (100 pickup and delivery stops). In general, the scheduled time segments of the problems have a length of two or three hours. The operated tours were planned by the standard software product in spring/winter 2014 and in winter 2015. Since the standard software does not optimize the number of couriers, the number of required vehicles is estimated in advance. Next, multiple runs with a different number of vehicles in the estimated range are performed in parallel. The result with the fewest number of unserved orders is returned. In contrast, the multiagent system allows explicitly minimizing the number of vehicles in a single run. Therefore, the system starts with an optimistic initial estimation of vehicles. Next, an *observer-agent* checks after a fixed number of negotiations if there are still unserved orders. If this is the case, a new vehicle agent is created. The vehicle, which the agent represents, is chosen from a sorted list which includes all available vehicles. The list was provided and sorted by tiramizoo. If the list is empty a provided prototype vehicle is created. This process is continued until the number of unserved orders is down to 0. Even if the initial number of vehicles is overestimated, the system implicitly reduces the number of couriers, because in most cases more efficient solutions include less vehicles. Thus, the agent representing a dispensable vehicle would (in general) not auction any order (cf. Section 4.3).[2] In the following investigation, the dispAgent software system is configured in such a way that order agents only accept the best proposal, instead of accepting the second best proposal in rare cases to guide the search in new directions and to avoid loops (cf. Section 4.3). This is because first experiments on real-world data revealed that orders with nearby pickup and delivery locations are mostly transported by the same vehicle and the investigated real-world problems are less restricted than the problems of the artificial benchmark sets of Solomon (1987) and Homberger and Gehring (2005). These are highly restricted, e.g., by time windows and the accuracy of the distance matrix as described in Chapter 5. Thus, in the investigated real-world problems, there is an increased probability that a solution is the best solution and not local optimum. The course-granular pre-processing (the initial clustering) is applied to accelerate

---

[2]Indeed, in very rare cases an additional agent could be removed if all possible reallocations (all permutations) would be investigated. However, as the problem is NP-hard (cf. Chapter 2), it is impractical for large-size problems. Thus, the computation could remain at a local minimum.

the search and to avoid synchronization conflicts particularly when large problem instances have to be solved (cf. Section 4.2). Since the preliminary runs revealed that the dispAgent software system solves the problem instances in reasonable time even if the course-granular planning is not applied, it is deactivated. Consequently, the system starts immediately with the negotiations and the fine-granular planning of the vehicle agents (cf. Section 4.3). The meta-information of performed tours, distance matrices, as well as the applied pre- and post-processing techniques were provided by tiramizoo.[3]

7.4.1 Problem Instances from Winter/Spring 2014

The first investigation focuses on 139 problem instances (tours operated in winter/spring 2014). Since the pre-processing did not shorten the delivery times for the computation of these tours (to ensure that vehicles return to the depot on time), the latest time for the vehicles to be back at the depot is also set to infinity in the test set for the dispAgent software system to allow an adequate comparison. The distance matrix which is used by the standard software product is transformed by Eq. 7.1. Unfortunately, a few operated tours are identified in which the vehicle did not return to the depot even if the test data for the dispAgent software system indicates that it should return to the depot. Thus, the following results are a conservative comparison from the perspective of the dispAgent approach.

The results show that the overall number of unserved orders is reduced from 25 to 0. In addition, the dispAgent software system reduces the total number of required vehicles by 13.96% and the number of stops is reduced by about 1.93%. The time-distance refers to the driving time of vehicles. Tours are evaluated by the driving time, because minimizing the time saves more cost than decreasing the distance. This is even more relevant in urban districts which are the main service areas of CEP services. The total time-distance increases by 0.03%. Figure 7.2 shows that in 13% of all simulated scenarios the

---

[3]Thus, also the time-distance matrix used by the dispAgent software system was computed and provided by tiramizoo.

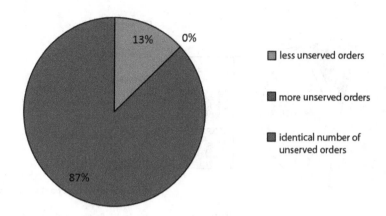

Figure 7.1: In all recomputed tour plans from spring/winter 2014, the dispAgent software system reduces the number of unserved orders to 0. In about 13% of all tours computed by the standard software product with pre- and post-processing at least one order remains unserved.

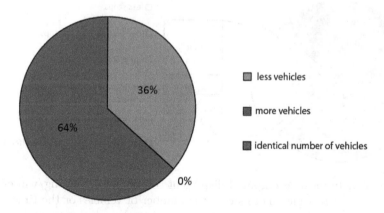

Figure 7.2: In about 36% of all recomputed tour plans from spring/winter 2014, the dispAgent software system reduces the required amount of vehicles compared to the solutions computed by the standard software product with pre- and post-processing.

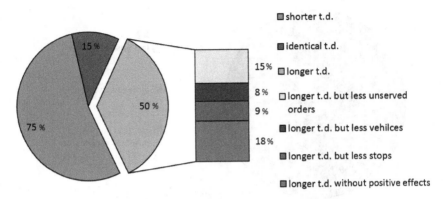

Figure 7.3: In most recomputed dispAgent solutions from spring/winter 2014 which have a longer time-distance (t.d.), the number of vehicles, unserved orders, or stops are decreased compared to the solutions computed by the standard software product with pre- and post-processing.

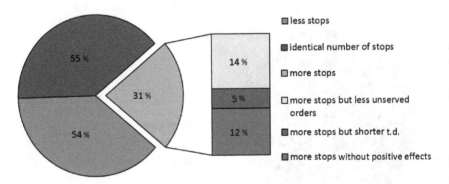

Figure 7.4: In most recomputed dispAgent solutions from spring/winter 2014 which more stops, the number of vehicles or the time distances are decreased compared to the solutions computed by the standard software product with pre- and post-processing.

number of unserved orders is reduced. Figure 7.2 depicts that in 36% of all recomputed tour plans, less vehicles are required. Figures 7.3–7.4 compare the time-distance and the amount of stops respectively. Both figures reveal that only in rare cases do the dispAgent solutions have a longer time-distance or more stops without any other positive effect such as the reduction of vehicles, of the amount of stops, or of the time-distance. The median computation time on a quad-core i7-4500 CPU 1.8 GHz with Windows 8 / 64 bit is 83 seconds while the maximum run-time for the largest tour plan is 1399 seconds.

7.4.2 Problem Instances from Winter 2015

The second investigation includes 62 tours (operated in winter 2015). For the computation of the operated tours the system applied all above mentioned pre- and post-processing procedures. The distance matrix which is used as input for the standard software product is scaled by Eq. 7.1 and additionally enlarged by 0.5%.[4] In the post-processing, all tours are recomputed based on the time-distances which are also directly used by the dispAgent software system. In the pre-processing for the standard software product, the service times are set to a single order at a stop. While in the tours from winter/spring 2014 the handling times are equal and defined at the orders' properties, the handling times in this investigation are computed by Eq. 7.2. Let k denote the number of orders with a pickup or delivery at a stop s. Then, the handling-time t_s of one order refers to the total time consumption at stop s. It is defined by

$$t_s = \begin{cases} 6 & \text{if } k \leq 4 \\ 8 & \text{if } k \geq 5 \text{ and } k \leq 10 \\ 12 & \text{otherwise.} \end{cases} \qquad (7.2)$$

All other orders which have to be handled at the same stop have a service time of 0. If orders with a pickup or delivery at this stop are

---

[4]Thus, also the traveling times are slightly longer, but increasing the time-distance is necessary to avoid an underestimation by Eq. 7.1 because of the arithmetric precision. The time-distance as well as all parameter values of Eq. 7.1 are only provided in hours with a precision of two.

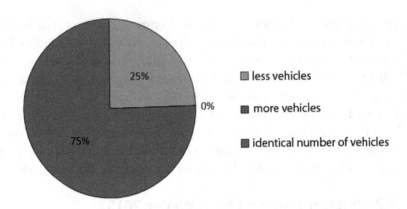

Figure 7.5: In about 25% of all recomputed tour plans from winter 2015, the dispAgent software system reduces the required amount of vehicles compared to the solutions computed by the standard software product with pre- and post-processing.

transported by different vehicles, the handling time is adjusted in post-processing. To this end, the post-processing results in delays at further customer stops as well as at the depot. In order to investigate the effects of post-processing, arbitrary tours were visualized and manually analyzed in cooperation with experts of tiramizoo. The observed delay of these tours is mostly some minutes (less than 10 minutes in general). In contrast, the dispAgent ensures that customer time windows are satisfied on time because no pre- and post-processing is necessary. Instead, the handling times computed by Eq. 7.2 as well as unmodified travel-times are explicitly considered in the agents decision making during tour construction. For an adequate comparison, the dispAgent system also allows a controlled delay of five minutes maximum for returning to the depot.

The results show that both systems compute solutions without unserved orders. However, the dispAgent software system reduces the total number of required vehicles by 8.21%. At the same time, the number of stops is reduced by about 0.95% and the total time-distance by 1.3%. Figure 7.5 depicts that in 25% of all recomputed tour plans

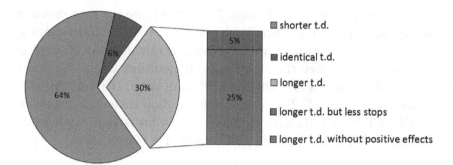

Figure 7.6: The most recomputed dispAgent solutions from winter 2015 have a shorter time-distance (t.d.) compared to the solutions computed by the standard software product with pre- and post-processing.

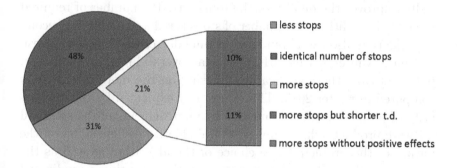

Figure 7.7: In multiple recomputed dispAgent solutions from winter 2015 which have more stops, the time-distance (t.d.) is decreased compared to the solutions computed by the standard software product with pre- and post-processing.

less vehicles are required. Figures 7.6–7.7 compare the time-distance and the amount of stops respectively. In most problems, the dispAgent solutions have a shorter time-distance. Also the number of stops is decreasing more often than increasing. In addition, Figure 7.7 shows that in about half of all solutions with more stops, the time-distance is decreasing instead. The median computation time on a quad-core i7-4500 CPU 1.8 GHz with Windows 8 / 64 bit is 77 seconds while the maximum time for the largest tour plan is 660 seconds.

7.4.3 Discussion

The comparison with the operated tours, which were computed by a standard software product with pre- and post-processing procedures, shows that the implementation of the dispAgent approach substantially improves the quality of the tours wrt. the number of required vehicles. In addition, the number of stops is reduced and in the second investigation also the required total time decreases. The first investigation in winter/spring 2014 reveals an overall higher optimization potential than the investigation focusing on the more recent tours computed in winter 2015. This is not only explained by the dissimilarity of problem instances but also by improvements of the standard software product, which applies several algorithms depending on the nature of the problem. The choice of the algorithm as well as the algorithms themselves were improved in the meantime as indicated by the company. In addition, the standard software product does not automatically optimize the number of couriers. Thus, the possibility remains that a solution with less vehicles is not found, because this setting was not investigated. In the meantime the estimations were improved on the basis of historical data and a higher degree of parallel running computations. This is also supported by the fact that in the investigation provided in Section 7.4.2 the number of unserved orders is reduced to 0 by both systems.

A detailed comparison of single tours in both investigations with experts from tiramizoo revealed that the tours computed by the dispAgent are more efficient than those computed by the standard software product. The standard software product requires pre- and post-processing procedures to satisfy the individual requirements of tiramizoo. Instead, the dispAgent approach considers all constraints directly during the optimization process and exploits the resulting potential for optimization. For instance, to ensure adequate handling times, the post-processing might increase handling-times which results in a delayed arrival at the depot (and also at further incoming goods departments of customers). Considering in advance a limited *delay* of five minutes for returning to the depot and considering this additional time directly in the tour planning process increases the flexibility and creates additional optimization potential which is exploited in the tour planning. This results in more efficient tours, ensures that all customer constraints are satisfied, and guarantees that a vehicle's delay is limited to five minutes. In contrast, the standard dispatching system with post-processing might cause even higher delays. Furthermore, the diapAgent directly uses a time-dependent distance matrix instead of scaling the distances to ensure reliable travel-times on the scaled distance with an average velocity. Thus, a negative impact caused even by small deviations resulting from arithmetic operations is avoided.

7.5 Qualitative Evaluation

As described above, the consideration of different vehicle types and properties such as individual capacities, velocities, and road-access in the planning and control is particularly relevant for CEP services providers. These requirements are considered by the vehicle agents of the multiagent system. During the decision-making process and tour planning, each agent values the proposed orders based on the available information, requirements, capabilities, and goals. For instance, to satisfy the access restrictions of a specific customer of tiramizoo, a vehicle agent only agrees upon a commitment if it completely fulfills the requirements of the order agent representing the transported goods of the customer. Optionally, an order agent sends *call-for-proposal* messages to preferred or trustworthy vehicle agents only.

Another advantage of the dispAgent approach is that the negotiation and communication protocols are designed for selfishly acting agents (cf. Section 4.3). Thus, the self-determination and autonomy of the competing actors is ensured, which is especially relevant in the domain of CEP services.

Nearby customers should be visited as soon as possible (without decreasing the efficiency). For example, in some cases the vehicle can visit customers on the outward run to another *far-away* customer instead of on the return run. This is also considered by the planning algorithms of the vehicle agents (cf. Section 4.4). Therefore, the cost calculation of the optimal branch-and-bound algorithm as well as the valuation in the rollout function of the NRPA-algorithm involve the total time span of orders being loaded onto the vehicle.

In addition, CEP services providers significantly profit from the dispAgent system in dynamic environments. The simulation of single scenarios (as well as the investigation presented in Chapter 6) proves that the dispAgent software system is able to consider dynamically incoming orders and vehicles. In case of unexpected events, it is also possible to remove certain vehicles. As described in Section 4.3, the system automatically reallocates orders and improves the solution quality with regard to the current order situation and environmental influences.

Another investigation of a real-world scenario focuses on cost-prediction of orders that have not yet been accepted to transport. The goal is to estimate the cost of a new incoming order in advance. For this purpose, a *temporal order agent* is created. This agent asks the vehicle agents for their prices to transport the new order. None of the received offers is accepted but the result is returned to the order management system. With this new information the order management system can decide which price should be suggested to the end customer. Thus, it can also be established whether a transport is profitable under the present circumstances. As the check is performed in less than 6 seconds on a quad-code i7-4500 CPU 1.8 GHz with Windows 8/ 64 bit laptop, the investigation confirms that the mechanism can be integrated in an online platform which interacts with the end-users.

Furthermore, the product range can be extended. The dispAgent approach allows reacting to dynamically changing customer requirements. For example, the order's delivery and pickup time windows or locations can change during the operations. The respective order agents take the new requirements into account and start a rescheduling. Thus, the system automatically updates the plans. If changing requirements cannot be performed, the dispAgent software system can inform the order management system.

7.6 Summary and Conclusion

This case study proves the applicability of the dispAgent system in one of the most complex and dynamic domains in logistics. At first, the special requirements of CEP services providers were identified and discussed. Next, the dispAgent approach was configured and extended to fulfill the requirements of the industrial partner tiramizoo. Finally, the evaluation was performed with real-world data of several days which was provided by tiramizoo. The unmodified data was taken from multiple service areas and contains all service requests of a complete schedule including the relevant performance measures of tours, which are computed by a commercial standard software product which implements, i.a., a large neighborhood search.

In real-world processes, the dispAgent clearly outperforms the standard software product which requires pre- and post-processing to satisfy company specific requirements. The results show substantial optimization potential in real-world operations by comparing the dispAgent tours to those computed by the commercial dispatching software product with pre- and post-processing. For instance, the numbers of both unserved orders and vehicles are reduced. Simultaneously, the number of stops and the required time is decreased. In addition, the system profits from its flexible, adaptive, and robust behavior in dynamic environments. Furthermore, it fulfills the domain-, customer-, and company-specific demands which are increasingly important in the domain of CEP services in order to extend the product range and provide high-quality services. As the dispAgent system returns a *kml-file*, which includes all computed tours, they can be visualized with GoogleEarth. Therefore, the correctness and applicability of randomly selected tours have been further discussed and successfully validated in detail with experts from tiramizoo. As a result, tiramizoo decided to incorporate the dispAgent software system in their operational processes and replace the standard software product by the dispAgent approach. The daily use in real-world operations further approves the applicability of the dispAgent approach and verifies the results of this case study.

8 Conclusion and Outlook

This thesis developed an autonomous multiagent-based approach for the optimization, analysis, planning, and control of logistics transport processes. The last chapter presents the overall summary and conclusion of the research and comes back to the initial research questions presented in Chapter 1 (cf. Section 8.1). As transport logistics in general is a wide interdisciplinary area, there remain aspects which cannot all be investigated in a single thesis. Therefore, the last part of the chapter focuses on further application domains (cf. Section 8.2) and provides research directions for future investigations (cf. Section 8.3).

8.1 Summary and Conclusion

It is the main goal of this thesis to develop an approach which allows handling rising challenges and increasing requirements in the transport sector. Both result from the integration of Industry 4.0 concepts in real-world processes (cf. Chapter 1). In order to achieve this goal, especially transport logistics must become more efficient, flexible, adaptive, reactive, and customized. Chapter 2 formally introduces general transport and routing problems with very common constraints and presents established, mostly centralized approaches to solve them. It turns out that they are inadequate to solve highly customized and constrained problems in vastly complex and dynamic domains. Chapter 3 concludes that the implementation of autonomous control is a way to overcome the identified weaknesses. This leads to the first research question:

1. How can autonomous control be implemented into current logistics processes so as to achieve smart transport logistics with low
investment?

It has been shown that multiagent systems are the technology of
choice to develop an approach which achieves the goals described in
the previous chapters. The reasons are as follows: Firstly, they provide
high flexibility, reactivity, proactivity, and robustness. Secondly, they
are suitable for comprehensible modeling real-world systems as agents
which are natural metaphors of physical objects and their correla-
tions as multiagent negotiations, coordination, and communication.
Therefore, the integration of customized constraints of changing areas
of application is possible with little effort. Finally, the decentralized
structure and emergent system behavior allow splitting up overall complex problems into smaller size problems with lower complexity. These
problems can be solved locally and often optimally. Subsequently,
Chapter 3 analyzes state-of-the-art multiagent-based systems in transport logistics in detail. It compares the approaches, identifies advantages and disadvantages, and finally discusses their suitability for
implementation in Industry 4.0 processes. Therefore, it is essential,
that agents run concurrently on high performance computers, e.g.,
in the cloud, to profit from the decentralized structure. This also
includes the support of conflict-free, highly parallelized multiagent
negotiations. In addition, the autonomy of selfishly acting agents must
be ensured. In the communication protocols and negotiations, the
agents must decide by themselves which information to reveal and they
must be able to keep confidential data private. Moreover, the overall
runtime performance of the multiagent system must be small enough
to satisfy real-world requirements and to adapt solutions online, which
is especially relevant in dynamic environments. This requires efficient
decision-making by the agents with high-performance algorithms.
Ultimately, for real-world application it is necessary to look at all
components of the multiagent system. This includes shortest-path
searches, which are completely neglected by other multiagent-based
approaches.

To overcome the identified weaknesses of state-of-the-art approaches and to satisfy the requirements for the implementation of autonomous control in transport logistics, the next research questions were answered:

2. Which communication and negotiation mechanisms allow for multi-agent-based autonomous control in transport logistics?

3. How must the decision-making processes of autonomously acting agents be designed to satisfy the requirements of transport service providers and customers in real-time?

Chapter 4 presents an inherent autonomous multiagent-based approach for optimizing real-world processes which satisfies all mentioned requirements. Established as well as recently developed concepts and components are taken from state-of-the-art solutions, adapted, extended, and enriched with new kinds of algorithms. At the beginning, a static pre-processing is applied to analyze the current order situation and compute an initial solution. Next, the agents continue optimizing the solution by stable negotiation mechanisms designed for selfishly acting, rational agents which can in particular be applied for dynamic negotiations to improve allocations continuously. It is not possible for an agent to manipulate the outcome of negotiations for its own purposes. Thus, the multiagent system facilitates the integration of participants with conflicting interests in a cooperative setting. The protocols allow for a high degree of parallel negotiations. Mechanisms are integrated to reduce the likelihood of conflicts in these negotiations. Moreover, an optimistic synchronization mechanism ensures consistency and integrity of communication. The overall communication effort is reduced to a minimum number of required messages. In addition, the protocols ensure the protection of privacy because all information remains with the agents themselves. To satisfy the requirements of the industrial partners, every negotiation can be stopped at any time and a consistent solution is returned.

Moreover, Chapter 4 presents high-performance algorithms which enable the agents to make optimal decisions as well as decisions with near to optimal results within a short time online. While the first mentioned is an efficient implementation of a customized branch-and-bound algorithm, the latter one is a novel approach to apply Nested Rollout Policy Adaptation for pickup and delivery routing problems. Both algorithms consider customized requirements of our industrial partners and can be further adapted with little effort to consider additional constraints.

In order to develop an inherent approach for real-world requirements, the topic of shortest-path computation is also considered. By analyzing the impact of shortest-path searches on the overall system runtime behavior, it turns out that the application of high performance algorithms is essential for the multiagent-based approach. Therefore, also a suitable and high-performance agent architecture is developed which reduces the runtime significantly.

The last research question is:

4. How large is the optimization potential of industrial partners as a result of implementing multiagent-based autonomous control in their current transport processes?

Firstly, since there are no adequate benchmark sets for dynamic transport scenarios, Chapter 5 investigates the solution quality with the established static benchmarks sets of Solomon (1987) and Homberger and Gehring (2005). Although the best approaches which compute state-of-the-art solutions are clearly dominated by Operations Research (OR) solvers, the multiagent-based approach computes suitable solutions for these artificial problems (as mentioned above this is remarkable because even static OR solvers have problems finding best-known solutions although they cannot be applied in dynamic real-world environments). Compared to the solution quality of solutions computed with other multiagent systems, the multiagent-based approach is at least competitive with or even outperforms these systems.

Beside their evaluation with established benchmark sets for static VRPs, the performance of the approach was tested in two of the most dynamic and complex domains in transport logistics, namely groupage traffic (cf. Chapter 6) and courier, express, and parcel services (cf. Chapter 7). Therefore, two case studies were performed in cooperation with our industrial partners, the Bremen office of Hellmann Worldwide Logistics and tiramizoo GmbH. The first is an office of one of the largest transport service provider in the world, the second a leading company for same-day delivery. The case studies show that the approach handles complex as well as dynamic scenarios due to the adaptive and reactive behavior of the intelligent agents as well as to the agents' high performance decision-making which considers all relevant domain-dependent constraints. It further proved that the approach can easily be customized to different domains in logistics.

In groupage traffic, the overall efficiency is increased significantly by reducing the number of expensive external freight carriers called at short notice. Beside the consideration of external events in the dynamic environment, the agents include premium service priorities in their decision-making process. As a result, conventional orders, which might be delivered or picked up on following days, are automatically shifted to next day schedules if an immediate premium service (which also might come in during operations) can be handled instead without calling an extra (external) freight carrier.

In the second case-study in courier, express, and parcel services, the computed tours were compared to schedules of a commercial standard software product. The results show that the tours of the multiagent-based approach clearly outperform those of the standard software product. This is due to improved consideration of all domain-dependent and relevant constraints directly in the planning process. Consequently, the tiramizoo GmbH decided to incorporate the developed approach in their operational processes for daily tour planning. The applicability of the multiagent-based approach is, therefore, further proven in real-world application.

In both case studies, the results, conclusions, and tours were additionally discussed, and successfully validated with experienced experts of our industrial partners.

8.2 Future Application Domains

Technological advances enable logistics services providers to exploit further optimization potential and to improve the synchronization between processes. On the one hand, breakdowns, delays, and rescheduling of production processes are insufficiently considered by logistics systems. As production plants are not connected to logistics systems, no information about the state of the production facility is directly considered in the planning and control of the logistics processes in real-time. Thus, the system cannot consider this information proactively and is not able to update the plan in case of relevant events. For example, if a component is required later, because the job flow has changed, the priority of the order can be reduced in the logistics system. Thus, the tour can be adapted and freed capacities can be used to process other jobs. As a result, calling cost intensive external freight carriers is avoided and the number of stops is reduced by consolidating orders later on.

On the other hand, breakdowns, delays, and rescheduling of logistics processes are also insufficiently considered by production systems. No information is directly processed by a production plant in real-time. For instance, if a consignment is delayed, other supply options can proactively be checked, or the job flow can be adapted to avoid downtimes of machines.

The presented multiagent-based approach can be applied to handle these scenarios from the perspective of logistics. Therefore, the IT systems of several participants have to be linked to each other and transmit their data in real time. Since IT systems in the real world are very heterogeneous, it is in most cases preferable to connect already applied IT systems via predefined interfaces, e.g., to a cloud computing platform. The modular multiagent-based structure allows mapping objects and their data to agents which in turn process the information and start the optimization. At the same time, another software component of a participating company may ask their assigned (owned) agents about their current status to monitor and analyze current processes and the status throughout the whole supply network.

Furthermore, the approach developed in this thesis increases customer service quality. It is possible to add new features demanded by customers, which will be increasingly important for future business (cf. Chapter 1). For instance, the multiagent system can be extended with little effort to handle dynamically changing delivery locations, time windows, priorities, amounts, etc. all of which also may change during operations.

In conclusion, the developed multiagent-based approach satisfies the challenging requirements in complex and dynamic environments. Therefore, it is especially designed to be applied in processes with a high volatility in which customized reactive behavior is essential, e.g., in Industry 4.0 processes or in smart urban logistics.

8.3 Future Research Directions

From a research perspective, there are several areas which could be further investigated to improve the developed approach as well as logistic transport systems in general. Firstly, it would be interesting to investigate subcomponents of the multiagent system. For instance, instead of a k-means clustering presented in Section 4.2, the effect of applying other pre-processing approaches could be analyzed to compute an initial (static) solution which is continuously improved by the multiagent system. In addition, the developed Nested Rollout Policy Adaptation algorithm in Section 4.4.2 only contains domain-dependent information in the rollout function. By changing the evaluation criteria of a solution in this function, the algorithm can be applied to other logistics domains as shown by Edelkamp, Gath, and Rohde (2014) for solving the packaging problem of a container. Furthermore, the shortest-path problem is a chapter of its own. While this thesis investigated the impact of shortest-path searches and provides a suitable modeling approach for the high-performance computation of shortest paths in multiagent systems, logistics in general will profit from future research on dynamically changing, time-dependent shortest-path search as shown by Delling and Wagner (2007), Ding, Yu, and Qin (2008), Batz et al. (2010), Geisberger et al. (2012), to name but a few.

It turned out that the decision-making of the agents is the most cost-intensive operation of the presented multiagent system. Nevertheless, it could be investigated how the underlying multiagent management system and message transfer affect the performance (provided by the JADE framework in the proposed implementation) in order to further decrease running times and accelerate the system. For instance, there has been research on more efficient message transfer between agents (Jander and Lamersdorf, 2013) and there are also alternative agent management systems such as Aimpulse Spectrum[1], which allows implementing highly scalable multiagent systems including millions of agents and reduces the running time compared to JADE (Lorig, Dammenhayn, Müller, and Timm, 2015).

In addition, it would be interesting to see how prediction and anticipation of future events such as delays at incoming goods departments, changing order situation, traffic, or the possibility of customers refusing to accept an order can be integrated in the tour planning to develop an autonomous learning dispatching system, which starts with the knowledge generated from historical (big) data. Thus, the proactive behavior of the system can be increased.

---

[1]For more information see: http://www.aimpulse.com (cited: 1.9.15).

Appendix

A Benchmark Evaluation

A.1 Solomon Benchmark Results with a Runtime of 15 Minutes

Table A.1: The VQ and DQ of randomized problem instances having shorter and more restricted time-windows of the Solomon benchmark set with 100 orders (runtime 15 minutes; rounded after the second decimal point).

Instance	Number of Vehicles	Distance in Units
R101	20	1755.84
R102	19	1605.98
R103	14	1425.90
R104	10	1167.64
R105	14	1564.08
R106	12	1365.97
R107	11	1297.07
R108	10	1163.37
R109	12	1417.14
R110	11	1323.86
R111	11	1196.70
R112	11	1272.15

Table A.2: The VQ and DQ of randomized problem instances having longer time and less restricted time-windows of the Solomon benchmark set with 100 orders (runtime 15 minutes; rounded after the second decimal point).

Instance	Number of Vehicles	Distance in Units
R201	4	1828.23
R202	6	1973.55
R203	3	1631.98
R204	3	1210.77
R205	3	1393.42
R206	3	1437.35
R207	4	1561.84
R208	3	1201.52
R209	4	1642.25
R210	3	1554.39
R211	4	1587.90

Table A.3: The VQ and DQ of randomized and clustered problem instances having shorter and more restricted time-windows of the Solomon benchmark set with 100 orders (runtime 15 minutes; rounded after the second decimal point).

Instance	Number of Vehicles	Distance in Units
RC101	15	1766.88
RC102	13	1633.92
RC103	11	1364.07
RC104	11	1356.66
RC105	15	1653.50
RC106	12	1468.28
RC107	11	1423.60
RC108	11	1364.01

Table A.4: The VQ and DQ of randomized and clustered problem instances having longer and less restricted time-windows of the Solomon benchmark set with 100 orders (runtime 15 minutes; rounded after the second decimal point).

Instance	Number of Vehicles	Distance in Units
RC201	4	2113.33
RC202	5	2031.90
RC203	4	1648.06
RC204	3	1445.27
RC205	4	1914.12
RC206	4	1797.62
RC207	4	1954.52
RC208	4	1958.10

Table A.5: The VQ and DQ of clustered problem instances having shorter and more restricted time-windows of the Solomon benchmark set with 100 orders (runtime 15 minutes; rounded after the second decimal point).

Instance	Number of Vehicles	Distance in Units
C101	10	828.94
C102	10	1079.24
C103	10	1130.71
C104	10	1042.61
C105	10	906.69
C106	10	972.50
C107	10	1090.30
C108	10	1158.07
C109	10	1267.18

Table A.6: The VQ and DQ of clustered problem instances having longer and less restricted time-windows of the Solomon benchmark set with 100 orders (runtime 15 minutes; rounded after the second decimal point).

Instance	Number of Vehicles	Distance in Units
C201	3	683.54
C202	3	793.10
C203	4	1779.30
C204	4	1683.18
C205	3	928.61
C206	3	840.75
C207	3	709.40
C208	4	1529.47

A.2 Solomon Benchmark Results with a Runtime of 60 Minutes

Table A.7: The VQ and DQ of randomized problem instances having shorter and more restricted time-windows of the Solomon benchmark set with 100 orders (runtime 60 minutes; rounded after the second decimal point).

Instance	Number of Vehicles	Distance in Units
R101	20	1833.64
R102	18	1639.08
R103	14	1361.00
R104	10	1160.92
R105	14	1517.43
R106	12	1369.38
R107	11	1235.31
R108	10	1128.24
R109	12	1344.00
R110	11	1293.02
R111	11	1233.43
R112	10	1180.06

Table A.8: The VQ and DQ of randomized problem instances having longer time and less restricted time-windows of the Solomon benchmark set with 100 orders (runtime 60 minutes; rounded after the second decimal point).

Instance	Number of Vehicles	Distance in Units
R201	4	1608.44
R202	4	1833.54
R203	3	1431.36
R204	3	1343.68
R205	3	1527.82
R206	3	1284.27
R207	3	1327.67
R208	3	1131.81
R209	4	1643.98
R210	3	1340.57
R211	4	1470.44

Table A.9: The VQ and DQ of randomized and clustered problem instances having shorter and more restricted time-windows of the Solomon benchmark set with 100 orders (runtime 60 minutes; rounded after the second decimal point).

Instance	Number of Vehicles	Distance in Units
RC101	14	1742.29
RC102	12	1591.75
RC103	11	1422.42
RC104	10	1256.13
RC105	14	1651.97
RC106	12	1532.98
RC107	11	1343.17
RC108	10	1284.24

Table A.10: The VQ and DQ of randomized and clustered problem
instances having longer and less restricted time-windows
of the Solomon benchmark set with 100 orders (runtime 60
minutes; rounded after the second decimal point).

Instance	Number of Vehicles	Distance in Units
RC201	4	1886.24
RC202	4	1764.87
RC203	3	1600.67
RC204	3	1286.21
RC205	4	1989.35
RC206	4	1554.47
RC207	4	1637.47
RC208	3	1506.34

Table A.11: The VQ and DQ of clustered problem instances having
shorter and more restricted time-windows of the Solomon
benchmark set with 100 orders (runtime 60 minutes; rounded after the second decimal point).

Instance	Number of Vehicles	Distance in Units
C101	10	828.94
C102	10	1275.04
C103	10	1475.96
C104	10	1273.32
C105	10	917.30
C106	10	979.51
C107	11	1212.79
C108	10	1108.70
C109	10	1135.67

Table A.12: The VQ and DQ of clustered problem instances having
longer and less restricted time-windows of the Solomon
benchmark set with 100 orders (runtime 60 minutes; rounded after the second decimal point).

Instance	Number of Vehicles	Distance in Units
C201	4	675.81
C202	3	985.12
C203	4	809.55
C204	5	1553.09
C205	3	720.52
C206	3	890.02
C207	3	810.56
C208	3	1006.26

A.3 Homberger and Gehring Benchmark Results

Table A.13: The VQ and DQ of randomized problem instances having shorter and more restricted time-windows of the Homberger and Gehring benchmark set with 200 orders (runtime 60 minutes; rounded after the second decimal point).

Instance	Number of Vehicles	Distance in Units
C121	20	3111.26
C122	19	4460.05
C123	18	4153.72
C124	18	3850.17
C125	20	3057.29
C126	20	3018.17
C127	20	3371.79
C128	20	3521.84
C129	19	3706.08
C1210	29	3832.29

Table A.14: The VQ and DQ of randomized problem instances having longer time and less restricted time-windows of the Homberger and Gehring benchmark set with 200 orders (runtime 60 minutes; rounded after the second decimal point).

Instance	Number of Vehicles	Distance in Units
C221	7	3520.58
C222	7	3969.80
C223	7	3983.52
C224	8	5320.68
C225	7	3190.80
C226	7	3694.57
C227	6	2600.92
C228	7	4415.81
C229	7	4805.07
C2210	7	3910.13

Table A.15: The VQ and DQ of randomized and clustered problem instances having shorter and more restricted time-windows of the Homberger and Gehring benchmark set with 200 orders (runtime 60 minutes; rounded after the second decimal point).

Instance	Number of Vehicles	Distance in Units
R121	22	6626.63
R122	18	1470.44
R123	18	5697.80
R124	20	4986.41
R125	18	6008.79
R126	18	5482.07
R127	18	5309.33
R128	19	4810.98
R129	18	5417.33
R1210	18	5736.05

Table A.16: The VQ and DQ of randomized and clustered problem instances having longer and less restricted time-windows of the Homberger and Gehring benchmark set with 200 orders (runtime 60 minutes; rounded after the second decimal point).

Instance	Number of Vehicles	Distance in Units
R221	5	5963.46
R222	5	6094.85
R223	6	6487.09
R224	6	5316.89
R225	4	5631.67
R226	4	5496.16
R227	6	6435.68
R228	4	5470.66
R229	4	5919.70
R2210	4	5563.91

Table A.17: The VQ and DQ of clustered problem instances having shorter and more restricted time-windows of the Homberger and Gehring benchmark set with 200 orders (runtime 60 minutes; rounded after the second decimal point).

Instance	Number of Vehicles	Distance in Units
RC121	19	4746.61
RC122	19	4541.19
RC123	18	4378.82
RC124	21	4391.13
RC125	19	4260.03
RC126	19	4431.08
RC127	19	4478.10
RC128	18	4892.33
RC129	18	4765.69
RC1210	20	4982.59

Table A.18: The VQ and DQ of clustered problem instances having longer and less restricted time-windows of the Homberger and Gehring benchmark set with 200 orders (runtime 60 minutes; rounded after the second decimal point).

Instance	Number of Vehicles	Distance in Units
RC221	6	4510.37
RC222	6	1553.09
RC223	6	5286.60
RC224	5	4359.62
RC225	6	5934.20
RC226	6	5062.48
RC227	5	4884.72
RC228	5	4557.36
RC229	5	4512.08
RC2210	4	4151.59

B Groupage Traffic

B.1 Visualization of Tours

In order to clearly show and to analyze also interdependencies between computed tours of a day in groupage traffic, the following maps was created in cooperation with the *Aimpulse Intelligent Systems GmbH* which supported the project by the visualization of all tours of a day on a poster.

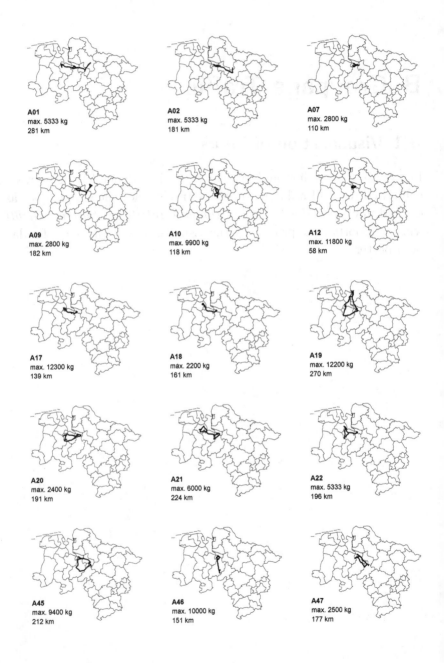

A01
max. 5333 kg
281 km

A02
max. 5333 kg
181 km

A07
max. 2800 kg
110 km

A09
max. 2800 kg
182 km

A10
max. 9900 kg
118 km

A12
max. 11800 kg
58 km

A17
max. 12300 kg
139 km

A18
max. 2200 kg
161 km

A19
max. 12200 kg
270 km

A20
max. 2400 kg
191 km

A21
max. 6000 kg
224 km

A22
max. 5333 kg
196 km

A45
max. 9400 kg
212 km

A46
max. 10000 kg
151 km

A47
max. 2500 kg
177 km

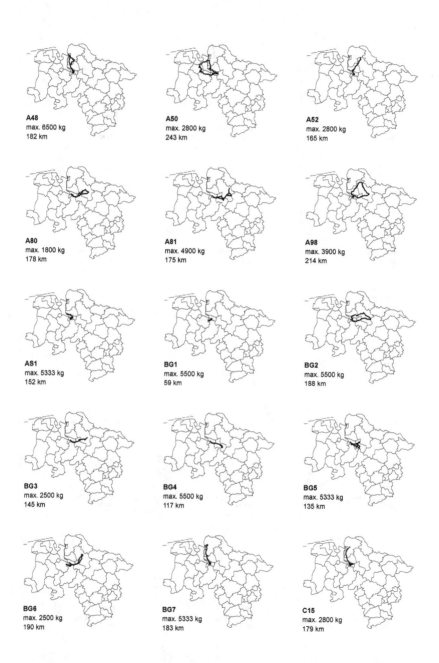

A48
max. 6500 kg
182 km

A50
max. 2800 kg
243 km

A52
max. 2800 kg
165 km

A80
max. 1800 kg
178 km

A81
max. 4900 kg
175 km

A98
max. 3900 kg
214 km

AS1
max. 5333 kg
152 km

BG1
max. 5500 kg
59 km

BG2
max. 5500 kg
188 km

BG3
max. 2500 kg
145 km

BG4
max. 5500 kg
117 km

BG5
max. 5333 kg
135 km

BG6
max. 2500 kg
190 km

BG7
max. 5333 kg
183 km

C15
max. 2800 kg
179 km

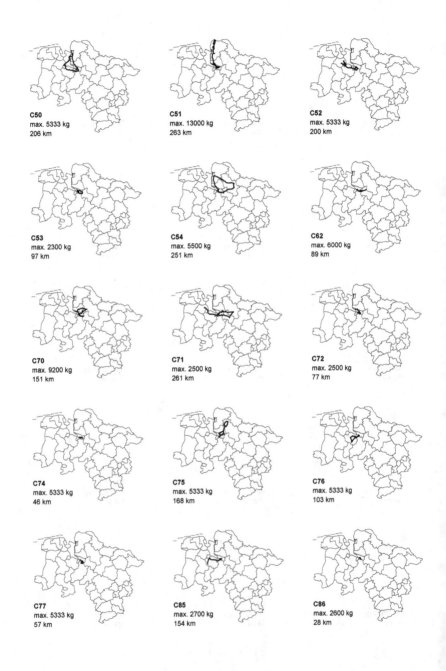

C50
max. 5333 kg
206 km

C51
max. 13000 kg
263 km

C52
max. 5333 kg
200 km

C53
max. 2300 kg
97 km

C54
max. 5500 kg
251 km

C62
max. 6000 kg
89 km

C70
max. 9200 kg
151 km

C71
max. 2500 kg
261 km

C72
max. 2500 kg
77 km

C74
max. 5333 kg
46 km

C75
max. 5333 kg
168 km

C76
max. 5333 kg
103 km

C77
max. 5333 kg
57 km

C85
max. 2700 kg
154 km

C86
max. 2600 kg
28 km

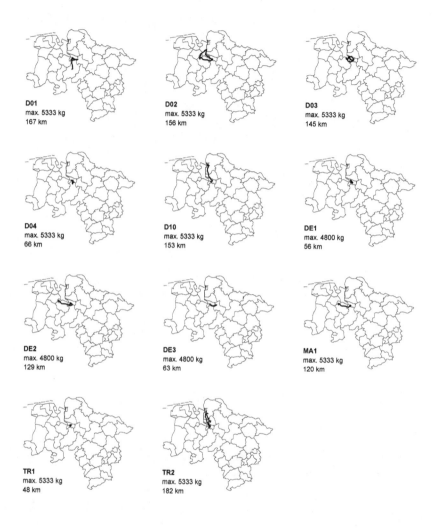

B.2 Expert Review with Hellmann Worldwide Logistics

aimpulse intelligent systems

Transferprojekt T8: Selbststeuernder Sammelgutverkehr
Expertenbewertung der automatisierten Dispositionsvorschläge

In der Niederlassung Bremen der Hellmann Worldwide Logistics GmbH & Co. KG (nachfolgend Hellmann) wurden in einem Forschungsprojekt (10/2011-03/2013) mit der Universität Bremen Einsatzmöglichkeiten und Optimierungspotentiale durch eine dynamische Transportdisposition untersucht. Im Projekt wurde ein System zur Unterstützung der Disposition entwickelt. Eine feingranulare Simulation hat ein Einsparpotenzial durch eine verbesserte Ressourceneffizienz und die Reduzierung von Sonderfahrten aufgezeigt.

In einem Gespräch mit ███████ als Experten für die Disposition wurde die Praxistauglichkeit der automatisch errechneten Touren und Routen bestätigt.

Teilnehmer des Expertengesprächs am 26.02.2013 bei Hellmann Bremen

████████████ ██████████████████, Hellmann-Niederlassung Bremen
Max Gath   Universität Bremen
Dr. Arne Schuldt   Aimpulse Intelligent Systems GmbH

Vorgehen

Als Grundlage für die Expertenbewertung dienten die für den 1. November 2011 errechneten Touren für insgesamt 56 Fahrzeuge. Für einen schnellen optischen Gesamteindruck diente ein Ausdruck sämtlicher Touren. Auf dieser Grundlage werden beispielhaft zahlreiche Touren im Detail untersucht. Für die Detailanalyse wurden sämtliche errechneten Touren mit Hilfe der Software Google Earth visualisiert. Neben dem Tourverlauf wurden relevante Kennzahlen und sämtliche Stopps dargestellt.

In der Simulation wurden die Aufträge der Niederlassung Bremen vom 1. November 2011 disponiert. Die Darstellung der Sendungsnummern auf der Karte ermöglichte es ███████ ██, die Ergebnisse der Disposition mit der Datenbank von Hellmann zu vergleichen.

Bewertung der Touren und Routen

███████ bewertet die automatisch berechneten Touren und Routen grundsätzlich als angemessen und durchführbar. Im Detail hat er die folgenden Verbesserungsvorschläge für einzelne Touren. Diese resultieren insbesondere aus dem Expertenwissen, das sich ██ im Rahmen seiner langjährigen Arbeit als Disponent angeeignet hat.

Länge der Touren. ███████ schätzt die Länge der vorgeschlagenen Touren als angemessen ein. Eine Ausnahme scheint ihm Tour C71 mit 14 Stopps und 260 km Länge zu sein. Er weist darauf hin, dass diese Tour zwar möglich ist, zur Einhaltung der Arbeitszeitrichtlinien aber darauf zu achten ist, dass ein Fahrer nicht an mehreren Tagen derartige Touren bekommt.

Stand- und Wartezeiten. Die Stand- und Wartezeiten in der Simulation sind laut ███████ ███████ angemessen. Allerdings weist er darauf hin, dass Stopps bei einzelnen Unternehmen (etwa bei der BLG) aus seiner Erfahrung länger dauern. Dies sollte der Disponent im Anschluss an die automatische Planung prüfen.

Be- und Entladung. ▮▮▮▮▮▮ weist darauf hin, dass insbesondere großvolumige Sendungen wie beispielsweise Fahrräder erst abgeholt werden können, wenn sämtliche Sendungen verteilt sind. In der vorgeschlagenen Tour MA1 würde er daher eine leichte Änderung der Reihenfolge der Stopps vornehmen.

Dynamische Disposition. Im Laufe des Tages eingegangene Abholaufträge werden in der automatischen Disposition dynamisch berücksichtigt. Beispielsweise ist Tour C85 frühzeitig mit der Bearbeitung von Sendungen in Bremen fertig und holt im Anschluss noch eine Sendung in der Nähe von Großenkneten ab. ▮▮▮▮▮▮ hält dies für eine angemessene Lösung, weist aber darauf hin, dass auch hier aufgrund der Arbeitszeitrichtlinien darauf zu achten ist, dass solche Touren verschiedenen Fahrern zugeteilt werden.

Eignung von Streckenabschnitten. Nach der Erfahrung von ▮▮▮▮▮▮ sind einige Streckenabschnitte, insbesondere in der Gegend um Hude und Ganderkesee, nicht für Fahrzeuge mit Anhänger geeignet. Um derartige Beschränkungen zu berücksichtigen, könnte dies in der Karte hinterlegt werden. Alternativ müsste der Disponent dies in automatisch vorgeschlagenen Touren noch einmal prüfen.

Belieferung von Privatkunden. ▮▮▮▮▮▮ weist darauf hin, dass Privatkunden (bspw. in Tour D04) am besten im Zeitfenster zwischen 10 und 14 Uhr beliefert werden. Dies sollte möglichst bei der automatischen Planung berücksichtigt werden – oder vom Disponenten in den vorgeschlagenen Touren durch Änderung der Reihenfolge der Stopps eingepflegt werden.

Zollausschlussgebiet. Einer der Stopps von Tour D10 liegt im Zollausschlussgebiet Fischereihafen in Bremerhaven. Da für ein- und ausgeführte Sendungen ein Dokument zu erstellen ist, empfiehlt ▮▮▮▮▮▮, diesen Stopp als letzten anzufahren. Auf der Grundlage der vorgeschlagenen Tour würde er die Reihenfolge der Stopps entsprechend anpassen.

Bei den Anmerkungen von ▮▮▮▮▮▮ ist festzustellen, dass diese in der Regel nicht die Zusammenstellung von Touren als Ganzes, sondern meist nur die Abfolge von Stopps im Detail betreffen. Das bedeutet, dass das Verfahren der dynamischen Disposition zur Unterstützung der Disponenten eingesetzt werden kann. Als automatisiertes System können in der Planung so mehr Bedingungen zuverlässig berücksichtigt werden, um das im Projekt identifizierte Einsparpotenzial zu heben. Auf der Grundlage einer solchen automatisierten Planung kann der Disponent dann seine Stärke ausspielen und mit Hilfe seines Expertenwissens die Planung punktuell verbessern.

Bibliography

Aarts, E., J. Korst, and P. van Laarhoven (1988). A quantitative analysis of the simulated annealing algorithm: A case study for the traveling salesman problem. *Journal of Statistical Physics 50*(1-2), 187–206.

Abraham, I., D. Delling, A. Goldberg, and R. Werneck (2011). A Hub-Based Labeling Algorithm for Shortest Paths in Road Networks. In P. Pardalos and S. Rebennack (Eds.), *Experimental Algorithms*, Volume 6630 of *Lecture Notes in Computer Science*, pp. 230–241. Springer Berlin Heidelberg.

Abraham, I., D. Delling, A. Goldberg, and R. Werneck (2012). Hierarchical hub labelings for shortest paths. In L. Epstein and P. Ferragina (Eds.), *Algorithms – ESA 2012*, Volume 7501 of *Lecture Notes in Computer Science*, pp. 24–35. Springer Berlin Heidelberg.

Ahlbrecht, T., J. Dix, M. Köster, P. Kraus, and J. P. Müller (2014). A Scalable Runtime Platform for Multiagent-Based Simulation. Technical report.

Ahuja, R. K., K. Mehlhorn, J. Orlin, and R. E. Tarjan (1990). Faster Algorithms for the Shortest Path Problem. *J. ACM 37*(2), 213–223.

Applegate, D. L., R. E. Bixby, V. Chvatal, and W. J. Cook (2006). *The Traveling Salesman Problem: A Computational Study*. Princeton, NJ, USA: Princeton University Press.

Austin, J. L. (1975). *How to do Things with Words* (2ed ed.). Oxford, UK: Oxford University Press.

Azi, N., M. Gendreau, and J.-Y. Potvin (2010). An exact algorithm for a vehicle routing problem with time windows and multiple use of vehicles. *European Journal of Operational Research 202*(3), 756–763.

Balakrishnan, N. (1993). Simple Heuristics for the Vehicle Routeing Problem with Soft Time Windows. *The Journal of the Operational Research Society 44*(3), pp. 279–287.

Barán, B. and M. Schaerer (2003). A Multiobjective Ant Colony System for Vehicle Routing Problem with Time Windows. In *Proceedings of the Twenty-First IASETED International Conference on Applied Informatics*, Innsbruck, Austria, pp. 97–102.

Barbucha, D. and P. Jędrzejowicz (2008). Multi-agent platform for solving the dynamic vehicle routing problem. In *Proceedings of the Eleventh International IEEE Conference on Intelligent Transportation Systems*, pp. 517–522.

Batz, G. V., R. Geisberger, S. Neubauer, and P. Sanders (2010). Time-Dependent Contraction Hierarchies and Approximation. In P. Festa (Ed.), *Experimental Algorithms*, Volume 6049 of *Lecture Notes in Computer Science*, pp. 166–177. Springer Berlin Heidelberg.

Bauer, R., T. Columbus, B. Katz, M. Krug, and D. Wagner (2010). Preprocessing Speed-Up Techniques Is Hard. In T. Calamoneri and J. Diaz (Eds.), *Algorithms and Complexity*, Volume 6078 of *Lecture Notes in Computer Science*, pp. 359–370. Springer Berlin Heidelberg.

Beaudry, A., G. Laporte, T. Melo, and S. Nickel (2010). Dynamic transportation of patients in hospitals. *OR Spectrum 32*(1), 77–107.

Bellifemine, F., G. Caire, and D. Greenwood (2007). *Developing Multi-Agent Systems with JADE*. Chichester, UK: John Wiley & Sons.

Bent, R. and P. V. Hentenryck (2006). A two-stage hybrid algorithm for pickup and delivery vehicle routing problems with time windows. *Computers & Operations Research 33*(4), 875–893.

Bent, R. W. and P. V. Hentenryck (2004). Scenario-Based Planning for Partially Dynamic Vehicle Routing with Stochastic Customers. *Operations Research 52*(6), 977–987.

Berbeglia, G., J.-F. Cordeau, and G. Laporte (2010). Dynamic pickup and delivery problems. *European Journal of Operational Research 202*(1), 8–15.

Berger, J. and M. Barkaoui (2004). A parallel hybrid genetic algorithm for the vehicle routing problem with time windows. *Computers & Operations Research 31*(12), 2037–2053.

Bjarnason, R., A. Fern, and P. Tadepalli (2009). Lower Bounding Klondike Solitaire with Monte-Carlo Planning. In *Proceedings of the Nineteenth International Conference on Automated Planning and Scheduling (ICAPS)*, pp. 26–33.

Branke, J., M. Middendorf, G. Noeth, and M. Dessouky (2005). Waiting Strategies for Dynamic Vehicle Routing. *Transportation Science 39*(3), 298–312.

Braubach, L., A. Pokahr, and W. Lamersdorf (2005). Jadex: A BDI Agent System Combining Middleware and Reasoning. In R. Unland, M. Calisti, and M. Klusch (Eds.), *Software Agent-Based Applications, Platforms and Development Kits*, pp. 143–168. Birkhäuser-Verlag Basel-Boston-Berlin.

Braubach, L., A. Pokahr, and W. Lamersdorf (2013a). Jadex Active Components: A Unified Execution Infrastructure for Agents and Workflows. In C. Enăchescu, F. G. Filip, and B. Iantovic (Eds.), *Advanced Computational Technologies*, pp. 128–149. Romanian Academy Publishing House Bucharest. (to appear).

Braubach, L., A. Pokahr, and W. Lamersdorf (2013b). Negotiation-based Patient Scheduling in Hospitals – Reengineering Message-based Interactions with Services. In B. Iantovics and R. Koutchev (Eds.), *Advanced Intelligent Computational Technologies and Decision Support Systems*, Number 486 in Studies in Computational Intelligence, pp. 107–121. Springer International Publishing Switzerland.

Braubach, L., A. Pokahr, W. Lamersdorf, K.-H. Krempels, and P.-O. Woelk (2006). A Generic Time Management Service for Distributed Multi-Agent Systems. *Applied Artificial Intelligence 20*(2-4), 229–249.

Bräysy, O., W. Dullaert, and M. Gendreau (2004). Evolutionary Algorithms for the Vehicle Routing Problem with Time Windows. *Journal of Heuristics 10*(6), 587–611.

Bräysy, O. and M. Gendreau (2005a). Vehicle Routing Problem with Time Windows, Part I: Route Construction and Local Search Algorithms. *Transportation Science 39*(1), 104–118.

Bräysy, O. and M. Gendreau (2005b). Vehicle Routing Problem with Time Windows, Part II: Metaheuristics. *Transportation Science 39*(1), 119–139.

Brooks, R. (1986). A Robust Layered Control System for a Mobile Robot. *IEEE Journal of Robotics and Automation 2*(1), 14–23.

Browne, C., E. Powley, D. Whitehouse, S. Lucas, P. Cowling, P. Rohlfshagen, S. Tavener, D. Perez, S. Samothrakis, and S. Colton (2012). A Survey of Monte Carlo Tree Search Methods. *IEEE Transactions on Computational Intelligence and AI in Games 4*(1), 1–43.

Bürckert, H.-J., K. Fischer, and G. Vierke (2000). Holonic Transport Scheduling with TeleTruck. *Applied Artificial Intelligence 14*(7), 697–725.

Cardeneo, A. (2008). Kurier-, Express- und Paketdienste. In D. Arnold, H. Isermann, A. Kuhn, and H. Tempelmeier (Eds.), *Handbuch Logistik*, Volume 3, pp. 782–788. Springer Berlin Heidelberg.

Cazenave, T. (2009). Nested Monte-Carlo Search. In *Proceedings of the Twenty-First International Joint Conference on Artifical Intelligence (IJCAI)*, San Francisco, CA, USA, pp. 456–461. Morgan Kaufmann Publishers Inc.

Cazenave, T. and F. Teytaud (2012). Application of the Nested Rollout Policy Adaptation Algorithm to the Traveling Salesman Problem with Time Windows. In Y. Hamadi and M. Schoenauer (Eds.), *Learning and Intelligent Optimization*, Volume 7219 of *Lecture Notes in Computer Science*, pp. 42–54. Springer Berlin Heidelberg.

Chen, Z.-L. and H. Xu (2006). Dynamic Column Generation for Dynamic Vehicle Routing with Time Windows. *Transportation Science 40*(1), 74–88.

Christofides, N. (1976). Worst-Case Analysis of a New Heuristic for the Travelling Salesman Problem. Technical Report 388, Graduate School of Industrial Administration, Carnegie-Mellon University.

Clarke, G. and J. W. Wright (1964). Scheduling of Vehicles from a Central Depot to a Number of Delivery Points. *Operations Research 12*, 568–581.

Cook, W. (2012). *In Pursuit of the Traveling Salesman: Mathematics at the Limits of Computation*. Princeton, NY, USA: Princeton University Press.

Cordeau, J.-F., M. Iori, G. Laporte, and J. J. Salazar González (2010). A branch-and-cut algorithm for the pickup and delivery traveling salesman problem with LIFO loading. *Networks 55*(1), 46–59.

Cordeau, J.-F. and G. Laporte (2003). A tabu search heuristic for the static multi-vehicle dial-a-ride problem. *Transportation Research Part B: Methodological 37*(6), 579 –594.

Cordeau, J.-F. and G. Laporte (2007). The dial-a-ride problem: models and algorithms. *Annals of Operations Research 153*(1), 29–46.

Créput, J.-C. and A. Koukam (2009). A memetic neural network for the euclidean traveling salesman problem. *Neurocomputing 72*(4), 1250–1264.

Dantzig, G. B. and J. H. Ramser (1959). The Truck Dispatching Problem. *Management Science 6*(1), 80–91.

Davidsson, P., L. Henesey, L. Ramstedt, J. Törnquist, and F. Wernstedt (2005). An Analysis of Agent-based Approaches to Transport Logistics. *Transportation Research Part C: Emerging Technologies 13*(4), 255 – 271.

Delfmann, W. and F. Jaekel (2012). Individuell bewegen - Das Internet der Dinge und Dienste. In W. Delfmann and T. Wimmer (Eds.), *Coordinated Autonomous Systems*, pp. 5–27. DVV Media Group.

Delling, D. and D. Wagner (2007). Landmark-Based Routing in Dynamic Graphs. In C. Demetrescu (Ed.), *Experimental Algorithms*, Volume 4525 of *Lecture Notes in Computer Science*, pp. 52–65. Springer Berlin Heidelberg.

Deutsche Post AG (2012). Einkaufen 4.0 - Der Einfluss von E-Commerce auf Lebensqualität und Einkaufsverhalten. http://www.dhl.de/content/dam/dhlde/downloads/paket/gk/studien/120209_dpdhl_studie_einkaufen4.0.pdf(cited: 1.9.15).

Dijkstra, E. (1959). A note on two problems in connexion with graphs. *Numerische Mathematik 1*(1), 269–271.

Ding, B., J. X. Yu, and L. Qin (2008). Finding Time-dependent Shortest Paths over Large Graphs. In *Proceedings of the Eleventh*

International Conference on Extending Database Technology: Advances in Database Technology, EDBT, New York, NY, USA, pp. 205–216. ACM.

Dorer, K. and M. Calisti (2005). An Adaptive Solution to Dynamic Transport Optimization. In *Proceedings of the Fourth International Joint Conference on Autonomous and Multiagent Systems (AAMAS)*, New York, NY, USA, pp. 45–51. ACM.

Dorigo, M. and L. Gambardella (1997). Ant colony system: a cooperative learning approach to the traveling salesman problem. *IEEE Transactions on Evolutionary Computation 1*(1), 53–66.

Dullaert, W. and O. Bräysy (2003). Routing relatively few customers per route. *Top 11*(2), 325–336.

Dumas, Y., J. Desrosiers, E. Gelinas, and M. M. (1995). An Optimal Algorithm for the Traveling Salesman Problem with Time Windows. *Operations Research 43*(2), 367–371.

Edelkamp, S. and M. Gath (2013). Optimal Decision Making in Agent-Based Autonomous Groupage Traffic. In J. Filipe and A. L. N. Fred (Eds.), *Proceedings of the Fifth International Conference on Agents and Artificial Intelligence (ICAART)*, Volume 1, pp. 248–254. SciTePress.

Edelkamp, S. and M. Gath (2014). Solving Single-Vehicle Pickup-and-Delivery Problems with Time Windows and Capacity Constraints using Nested Monte-Carlo Search. In B. Duval, J. van den Herik, S. Loiseau, and J. Filipe (Eds.), *Proceedings of the Sixth International Conference on Agents and Artificial Intelligence (ICAART)*, Volume 1, pp. 22–33. SciTePress.

Edelkamp, S., M. Gath, T. Cazenave, and F. Teytaud (2013). Algorithm and knowledge engineering for the TSPTW problem. In *IEEE Symposium on Computational Intelligence in Scheduling (SCIS)*, pp. 44–51.

Edelkamp, S., M. Gath, C. Greulich, M. Humann, O. Herzog, and M. Lawo (2015). Monte-Carlo Tree Search for Production and Logistics. In U. Clausen, H. Friedrich, C. Thaller, and C. Geiger (Eds.), *Proceedings of the Second Interdisciplinary Conference on Production, Logistics and Traffic (ICPLT)*, pp. 427–437. Springer International Publishing Switzerland.

Edelkamp, S., M. Gath, C. Greulich, M. Humann, and T. Warden (2014). PlaSMA multiagent simulation. Last-mile connectivity Bangalore. In O. Herzog (Ed.), *German Indian Partnership for IT Systems*, pp. 129–185. München/Berlin: acatech.

Edelkamp, S., M. Gath, and M. Rohde (2014). Monte-carlo tree search for 3d packing with object orientation. In C. Lutz and M. Thielscher (Eds.), *KI 2014: Advances in Artificial Intelligence*, Volume 8736 of *Lecture Notes in Computer Science*, pp. 285–296. Springer International Publishing.

Een, N., A. Mishchenko, and N. Srensson (2007). Applying Logic Synthesis for Speeding Up SAT. In J. a. Marques-Silva and K. Sakallah (Eds.), *Theory and Applications of Satisfiability Testing – SAT 2007*, Volume 4501 of *Lecture Notes in Computer Science*, pp. 272–286. Springer Berlin Heidelberg.

Esser, K. and J. Kurte (2014). Wirtschaftliche Bedeutung der KEP-Branche. Technical report, Bundesverband Paket und Express Logistik e.V. (BIEK). http://www.biek.de/index.php/studien.html (cited: 1.9.15).

Ferber, J. (1999). *Multi-Agent Systems: An Introduction to Distributed Artificial Intelligence*, Volume 1. Boston, MA, USA: Addison-Wesley Reading.

Fiechter, C.-N. (1994). A parallel tabu search algorithm for large traveling salesman problems. *Discrete Applied Mathematics 51*(3), 243–267.

Fischer, K., J. P. Müller, and M. Pischel (1996). Cooperative Transportation Scheduling: An Application Domain for DAI. *Journal of Applied Artificial Intelligence 10*(1), 1–33.

Fischetti, M., J. J. Salazar González, and P. Toth (1997). A Branch-and-Cut Algorithm for the Symmetric Generalized Traveling Salesman Problem. *Operations Research 45*(3), 378–394.

Fleisch, E. and F. Mattern (Eds.) (2005). *Das Internet der Dinge – Ubiquitous Computing und RFID in der Praxis: Visionen, Technologien, Anwendungen, Handlungsanleitungen*, Volume 1. Springer Berlin Heidelberg.

Fleischmann, B. (2008). Grundlagen: Begriff der Logistik, logistische Systeme und Prozesse. In D. Arnold, H. Isermann, A. Kuhn, and H. Tempelmeier (Eds.), *Handbuch Logistik*, Volume 3, pp. 3–34. Springer Berlin Heidelberg.

Flood, M. M. (1956). The Traveling-Salesman Problem. *Operations Research 4*(1), 61–75.

Foundation for Intelligent Physical Agents (2002). FIPA Contract Net Interaction Protocol Specification. Standard No. SC00029H, Geneva, Switzerland.

Franklin, S. and A. Graesser (1997). Is it an agent, or just a program?: A taxonomy for autonomous agents. In J. Müller, M. J. Wooldridge, and N. R. Jennings (Eds.), *Intelligent Agents III Agent Theories, Architectures, and Languages*, Volume 1193 of *Lecture Notes in Computer Science*, pp. 21–35. Springer Berlin Heidelberg.

Frederickson, G., M. S. Hecht, and C. E. Kim (1976). Approximation algorithms for some routing problems. In *Seventeenth Annual Symposium on Foundations of Computer Science*, pp. 216–227.

Fujimoto, R. M. (2000). *Parallel and Distributed Simulation Systems*. John Wiley & Sons.

Gajpal, Y. and P. Abad (2009). An ant colony system (ACS) for vehicle routing problem with simultaneous delivery and pickup. *Computers & Operations Research 36*(12), 3215–3223.

Gambardella, L. M., Éric Taillard, and G. Agazzi (1999). MACS-VRPTW: A Multiple Colony System For Vehicle Routing Problems With Time Windows. In *New Ideas in Optimization*, pp. 63–76. McGraw-Hill.

Gamma, E., E. R. Johnson, R. Helm, and J. Vlissides (1994). *Design Patterns: Elements of Reusable Object-Oriented Software*. Pearson Education.

Garcia, B.-L., J.-Y. Potvin, and J.-M. Rousseau (1994). A parallel implementation of the Tabu search heuristic for vehicle routing problems with time window constraints. *Computers & Operations Research 21*(9), 1025–1033.

Garey, M. R., R. L. Graham, and D. S. Johnson (1976). Some NP-complete Geometric Problems. In *Proceedings of the Eighth Annual ACM Symposium on Theory of Computing (STOC)*, New York, NY, USA, pp. 10–22. ACM.

Gath, M., S. Edelkamp, and O. Herzog (2013a). Agent-based dispatching enables autonomous groupage traffic. *Journal of Artificial Intelligence and Soft Computing Research 3*(1), 27–40.

Gath, M., S. Edelkamp, and O. Herzog (2013b). Agent-based dispatching in groupage traffic. In *Proceedings of the IEEE Workshop on Computational Intelligence in Production and Logistics Systems (CIPLS)*, pp. 54–60.

Gath, M. and O. Herzog (2015). Intelligent Logistics 2.0. *German research magazine of the Deutsche Forschungsgemeinschaft 1*, 26–29.

Gath, M., O. Herzog, and S. Edelkamp (2013). Agent-based planning and control for groupage traffic. In *Proceedings of the Tenth In-*

ternational Conference and Expo on Emerging Technologies for a Smarter World (CEWIT, pp. 1–7.

Gath, M., O. Herzog, and S. Edelkamp (2014a). Agenten für eine optimierte Logistik. *RFID im Blick Sonderausgabe Industrie 4.0 und Logistik 4.0 aus Bremen*, 36–37.

Gath, M., O. Herzog, and S. Edelkamp (2014b). Autonomous and flexible multiagent systems enhance transport logistics. In *Proceedings of the Eleventh International Conference Expo on Emerging Technologies for a Smarter World (CEWIT)*, pp. 1–6.

Gath, M., O. Herzog, and S. Edelkamp (2016). Autonomous, Adaptive, and Self-Organized Multiagent Systems for the Optimization of Decentralized Industrial Processes. In J. Kolodziej, L. Correia, and J. M. Molina (Eds.), *Intelligent Agents in Data-Intensive Computing*, Volume 14 of *Studies in Big Data*, pp. 71–98. Springer International Publishing Switzerland.

Gath, M., O. Herzog, and M. Vaske (2015a). Concurrent and Distributed Shortest-Path Searches in Multiagent-based Transport Systems. Volume 9420 of *Transactions on Computational Collective Intelligence*, pp. 140–157. Springer International Publishing.

Gath, M., O. Herzog, and M. Vaske (2015b). Parallel shortest-path searches in multiagent-based simulations with plasma. In S. Loiseau, J. Filipe, B. Dval, and J. van den Herik (Eds.), *Proceedings of the Seventh International Conference on Agents and Artificial Intelligence (ICAART)*, Volume 1, pp. 15–21. SciTePress.

Gath, M., O. Herzog, and M. Vaske (2016). The Impact of Shortest Path Searches to Autonomous Transport Processes. In H. Kotzab, J. Pannek, and K.-D. Thoben (Eds.), *Dynamics in Logistics – Proceedings of the Fourth International Conference on Dynamics in Logistics (LDIC)*, Volume 1, pp. 79–90. Springer International Publishing.

Gath, M., T. Wagner, and O. Herzog (2012). Autonomous logistic processes of bike courier services using multiagent-based simulation. In M. Affenzeller, A. Bruzzone, F. D. Felice, D. D. R. Vilas, C. Frydman, M. Massei, and Y. Merkuryev (Eds.), *Proceedings of the Eleventh International Conference on Modeling and Applied Simulation*, pp. 134–142.

Gehrke, J. D., A. Schuldt, and S. Werner (2008). Quality Criteria for Multiagent-Based Simulations with Conservative Synchronisation. In M. Rabe (Ed.), *Proceedings of the Thirteenth ASIM Dedicated Conference on Simulation in Production and Logistics*, Stuttgart, pp. 545–554. Citeseer: Fraunhofer IRB Verlag.

Geisberger, E. and M. Broy (Eds.) (2012). *agendaCPS: Integrierte Forschungsagenda Cyber-Physical Systems*, Volume 1. Springer Berlin Heidelberg.

Geisberger, R., P. Sanders, D. Schultes, and D. Delling (2008). Contraction hierarchies: Faster and simpler hierarchical routing in road networks. In C. McGeoch (Ed.), *Experimental Algorithms*, Volume 5038 of *Lecture Notes in Computer Science*, pp. 319–333. Springer Berlin Heidelberg.

Geisberger, R., P. Sanders, D. Schultes, and C. Vetter (2012). Exact Routing in Large Road Networks Using Contraction Hierarchies. *Transportation Science 46*(3), 388–404.

Gendreau, M., F. Guertin, J.-Y. Potvin, and R. Séguin (2006). Neighborhood search heuristics for a dynamic vehicle dispatching problem with pick-ups and deliveries. *Transportation Research Part C: Emerging Technologies 14*(3), 157–174.

Gendreau, M., F. Guertin, J.-Y. Potvin, and E. Taillard (1999). Parallel Tabu Search for Real-Time Vehicle Routing and Dispatching. *Transportation Science 33*(4), 381–390.

Gendreau, M., G. Laporte, and R. Séguin (1996). A Tabu Search Heuristic for the Vehicle Routing Problem with Stochastic Demands and Customers. *Operations Research 44*(3), 469–477.

Gendreau, M., G. Laporte, and F. Semet (1998). A tabu search heuristic for the undirected selective travelling salesman problem. *European Journal of Operational Research 106*(23), 539–545.

Geng, X., Z. Chen, W. Yang, D. Shi, and K. Zhao (2011). Solving the traveling salesman problem based on an adaptive simulated annealing algorithm with greedy search. *Applied Soft Computing 11*(4), 3680–3689.

Ghiani, G., F. Guerriero, G. Laporte, and R. Musmanno (2003). Real-time vehicle routing: Solution concepts, algorithms and parallel computing strategies. *European Journal of Operational Research 151*(1), 1–11.

Gillett, B. and L. R. Miller (1974). A Heuristic Algorithm for the Vehicle-Dispatch Problem. *Operations Research 22*(2), 240–349.

Glaschenko, A., A. Ivaschenko, G. Rzevski, and P. Skobelev (2009). Multi-Agent Real Time Scheduling System for Taxi Companies. In *Proceedings of the Eighth International Conference on Autonomous Agents and Multiagent Systems (AAMAS)*, pp. 29–36.

Golden, B. L., S. Raghavan, and E. A. Wasil (2008). *The Vehicle Routing Problem: Latest Advances and New Challenges: latest advances and new challenges*, Volume 43. Springer Science & Business Media.

Grefenstette, J., R. Gopal, B. Rosmaita, and D. Van Gucht (1985). Genetic algorithms for the traveling salesman problem. In *Proceedings of the First International Conference on Genetic Algorithms and their Applications*, Mahwah, NJ, USA, pp. 160–168. Lawrence Erlbaum.

Greulich, C., S. Edelkamp, and M. Gath (2013). Agent-Based Multimodal Transport Planning in Dynamic Environments. In I. J. Timm and M. Thimm (Eds.), *KI 2013: Advances in Artificial Intelligence*, Volume 8077 of *Lecture Notes in Computer Science*, pp. 74–85. Springer Berlin Heidelberg.

Greulich, C., S. Edelkamp, M. Gath, T. Warden, M. Humann, O. Herzog, and T. G. Sitharam (2013). Enhanced Shortest Path Computation for Multiagent-based Intermodal Transport Planning in Dynamic Environments. In J. Filipe and A. L. N. Fred (Eds.), *Proceedings of the Fifth International Conference on Agents and Artificial Intelligence (ICAART)*, Volume 2, pp. 324–329. SciTePress.

Hanshar, F. T. and B. M. Ombuki-Berman (2007). Dynamic vehicle routing using genetic algorithms. *Applied Intelligence 27*(1), 89–99.

Hellmann Worldwide Logistics GmbH & Co. KG (2015). Nachhaltigkeits-Report 2014. http://www.hellmann.net/downloads/2365/Gesch%C3%A4ftsbericht%20und%20Nachhaltigkeitsreport%202014.pdf?1438779112.(cited: 1.9.15).

Helsgaun, K. (2009). General k-opt submoves for the lin–kernighan tsp heuristic. *Mathematical Programming Computation 1*(2-3), 119–163.

Hernández-Pérez, H. and J.-J. Salazar-González (2004). A branch-and-cut algorithm for a traveling salesman problem with pickup and delivery. *Discrete Applied Mathematics 145*(1), 126–139.

Himoff, J., G. Rzevski, and P. Skobelev (2006). Magenta Technology Multi-agent Logistics i-Scheduler for Road Transportation. In *Proceedings of the Fifth International Joint Conference on Autonomous Agents and Multiagent Systems (AAMAS)*, New York, NY, USA, pp. 1514–1521. ACM.

Himoff, J., P. Skobelev, and M. Wooldridge (2005). MAGENTA Technology: Multi-agent Systems for Industrial Logistics. In *Proceedings of the Fourth International Joint Conference on Autonomous Agents and Multiagent Systems (AAMAS)*, New York, NY, USA, pp. 60–66. ACM.

Homberger, J. and H. Gehring (2005). A two-phase hybrid metaheuristic for the vehicle routing problem with time windows. *European Journal of Operational Research 162*(1), 220–238.

Hosny, M. and C. Mumford (2010). The single vehicle pickup and delivery problem with time windows: Intelligent operators for heuristic and metaheuristic algorithms. *Journal of Heuristics 16*(3), 417–439.

Hosny, M. I. and C. L. Mumford (2007). Single vehicle pickup and delivery with time windows: made to measure genetic encoding and operators. In *Proceedings of the Genetic and Evolutionary Computation Conference (GECCO)*, pp. 2489–2496.

Hülsmann, M., B. Scholz-Reiter, and K. Windt (Eds.) (2011). *Autonomous Cooperation and Control in Logistics: Contributions and Limitations - Theoretical and Practical Perspectives*. Springer Berlin Heidelberg.

Ichoua, S., M. Gendreau, and J.-Y. Potvin (2006). Exploiting Knowledge About Future Demands for Real-Time Vehicle Dispatching. *Transportation Science 40*(2), 211–225.

Jaillet, P. (1988). A Priori Solution of a Traveling Salesman Problem in Which a Random Subset of the Customers Are Visited. *Operations Research 36*(6), 929–936.

Jander, K. and W. Lamersdorf (2013). Compact and Efficient Agent Messaging. In M. Dastani, J. Hübner, and B. Logan (Eds.), *Proceedings of the Tenth International Workshop on Programming Multi-Agent Systems (ProMAS)*, Volume Programming Multiagent Systems of *Lecture Notes in Computer Science*, pp. 108–122. Springer Berlin Heidelberg.

Jefferson, D. (1990). Virtual Time II: Storage Management in Conservative and Optimistic Systems. In *Proceedings of the Ninth Annual ACM Symposium on Principles of Distributed Computing (PODC)*, New York, NY, USA, pp. 75–89. ACM.

Jennings, N. and M. Wooldridge (1998). Applications of intelligent agents. In N. Jennings and M. Wooldridge (Eds.), *Agent Technology*, pp. 3–28. Springer Berlin Heidelberg.

Jennings, N. R. (2001, April). An Agent-based Approach for Building Complex Software Systems. *Communications of the ACM 44*(4), 35–41.

Jih, W.-R. and J.-J. Hsu (1999). Dynamic vehicle routing using hybrid genetic algorithms. In *Proceedings of the IEEE International Conference on Robotics and Automation*, Volume 1, pp. 453–458.

Jih, W.-R. and Y. Hsu (2004). A family competition genetic algorithm for the pickup and delivery problems with time window. *Bulletin of the College of Engineering 90*, 121–130.

Johnson, D. and L. McGeoch (2007). Experimental Analysis of Heuristics for the STSP. In G. Gutin and A. Punnen (Eds.), *The Traveling Salesman Problem and Its Variations*, Volume 12 of *Combinatorial Optimization*, pp. 369–443. Springer US.

Jonker, R. and T. Volgenant (1986). Improving the Hungarian assignment algorithm . *Operations Research Letters 5*(4), 171–175.

Jünemann, R. (1989). *Materialfluß und Logistik*. Springer Berlin Heidelberg.

Kagemann, H., W. Wahlster, and J. Helbig (2013). Recommendations for implementing the strategic initiative Industrie 4.0 – final report of the Industrie 4.0 working group. Technical report, acatech.

Kalina, P. and J. Vokřínek (2012a). Algorithm for vehicle routing problem with time windows based on agent negotiation. In *Proceedings of the Seventh Workshop on Agents In Traffic and Transportation (AAMAS)*.

Kalina, P. and J. Vokřínek (2012b). Parallel Solver for Vehicle Routing and Pickup and Delivery Problems with Time Windows based on Agent Negotiation. In *Proceedings of the IEEE International Conference on Systems, Man, and Cybernetics (SMC)*, pp. 1558–1563.

Karp, R. M. and J. M. Steele (1985). Probabilistic analysis of heuristics. In E. L. Lawler, J. K. Lenstra, A. H. G. R. Kan, and D. B. Shmoys (Eds.), *The Traveling Salesman Problem*, pp. 181–205. John Wiley & Sons.

Kergosien, Y., C. Lent, D. Piton, and J.-C. Billaut (2011). A tabu search heuristic for the dynamic transportation of patients between care units. *European Journal of Operational Research 214*(2), 442–452.

Kirn, S., O. Herzog, P. Lockemann, and O. Spaniol (Eds.) (2006). *Multiagent Engineering: Theory and Applications in Enterprises*. Springer Berlin Heidelberg.

Koenig, S., C. Tovey, M. Lagoudakis, V. Markakis, D. Kempe, P. Keskinocak, A. Kleywegt, A. Meyerson, and S. Jain (2006). The Power of Sequential Single-Item Auctions for Agent Coordination. In *Proceedings of the Twenty-First National Conference on Artificial intelligence (AAAI)*, Volume 2, Menlo Park, CA; Cambridge, MA, pp. 1625–1629. AAAI Press; MIT Press.

Kohout, R. and K. Erol (1999). In-Time Agent-Based Vehicle Routing with a Stochastic Improvement Heuristic. In *Proceedings of the Sixteenth Conference on Artificial Intelligence and the Eleventh on Innovative Applications of Artificial Intelligence (AAAI/IAAI 1999)*, pp. 864–869. AAAI Press.

Krishna, V. (2009). *Auction Theory* (2ed ed.). Academic Press.

Kuhn, H. W. (1955). The Hungarian method for the assignment problem. *Naval Research Logistics Quarterly 2*(1-2), 83–97.

Larsen, A., O. Madsen, and M. Solomon (2002). Partially Dynamic Vehicle Routing-Models and Algorithms. *The Journal of the Operational Research Society 53*(6), pp. 637–646.

Lee, H. L. (2002). Aligning supply chain strategies with product uncertainties. *California Management Review 44*(3), 105–119.

Lenstra, J. K. and A. H. G. R. Kan (1981). Complexity of vehicle routing and scheduling problems. *Networks 11*(2), 221–227.

Leong, H. W. and M. Liu (2006). A Multi-agent Algorithm for Vehicle Routing Problem with Time Window. In *Proceedings of the ACM Symposium on Applied Computing (SAC)*, New York, NY, USA, pp. 106–111. ACM.

Lewandowski, M., M. Gath, D. Werthmann, and M. Lawo (2013). Agent-based Control for Material Handling Systems in In-House Logistics – Towards Cyber-Physical Systems in In-House-Logistics Utilizing Real Size. In *Proceedings of the European Conference on Smart Objects, Systems and Technologies (SmartSysTech)*, pp. 1–5.

Li, H. and A. Lim (2001). A metaheuristic for the pickup and delivery problem with time windows. In *Proceedings of the Thirteenth International Conference on Tools with Artificial Intelligence*, pp. 160–167.

Lin, S. (1965a). Computer Solutions of the Traveling Salesman Problem. *Bell System Technical Journal 44*(10), 2245–2269.

Lin, S. (1965b, Dec). Computer Solutions of the Traveling Salesman Problem. *The Bell System Technical Journal 44*(10), 2245–2269.

Lin, S. and B. W. Kernighan (1973). An Effective Heuristic Algorithm for the Traveling-Salesman Problem. *Operations Research 21*(2), 498–516.

Lorig, F., N. Dammenhayn, D.-J. Müller, and I. J. Timm (2015). Measuring and Comparing Scalability of Agent-based Simulation Frameworks. In *Proceedings of the Thirteenth German Conference on Multiagent System Technologies (MATES)*, Lecture Notes in Coputer Science. Springer Berlin Heidelberg. (to appear).

Luby, M., A. Sinclair, and D. Zuckerman (1993). Optimal speedup of Las Vegas algorithms. *Information Processing Letters 47*(4), 173–180.

Lund, K., Oli, and J. M. Rygaard (1996). Vehicle Routing Problems with Varying Degrees of Dynamism. Technical report.

Macal, C. M. and M. J. North (2010). Tutorial on agent-based modelling and simulation. *Journal of Simulation 4*(3), 151–162.

MacQueen, J. et al. (1967). Some Methods for Classification and Analysis of Multivariate Observations. In *Proceedings of the Fifth Berkeley symposium on mathematical statistics and probability*, Volume 1, pp. 281–297. California, USA.

Máhr, T., J. Srour, M. de Weerdt, and R. Zuidwijk (2010). Can agents measure up? A comparative study of an agent-based and online optimization approach for a drayage problem with uncertainty . *Transportation Research Part C: Emerging Technologies 18*(1), 99–119. Information/Communication Technologies and Travel Behaviour Agents in Traffic and Transportation.

Malandraki, C. and R. B. Dial (1996). A restricted dynamic programming heuristic algorithm for the time dependent traveling salesman problem. *European Journal of Operational Research 90*(1), 45–55.

Manner-Romberg, Symanczyk, Ströh, Deecke, Bastron, and Marwig. (2009). Primärerhebung auf den Märkten für Kurier-, Express-

und Paketdienste. Technical report, MRU GmbH. www.bdkep.de/dokumente/2010-Studie.pdf(cited: 1.9.15).

Mes, M., M. van der Heijden, and A. van Harten (2007). Comparison of Agent-based Scheduling to look-ahead Heuristics for Real-Time Transportation Problems. *European Journal of Operational Research 181*(1), 59–75.

Miller, D. L. and J. F. Pekny (1991). Exact Solution of Large Asymmetric Traveling Salesman Problems. *Science 251*(4995), 754–761.

Mitrovic-Minic, S. and G. Laporte (2004). Waiting strategies for the dynamic pickup and delivery problem with time windows. *Transportation Research Part B: Methodological 38*(7), 635–655.

Montané, F. A. T. and R. D. G. ao (2006). A tabu search algorithm for the vehicle routing problem with simultaneous pick-up and delivery service. *Computers & Operations Research 33*(3), 595–619.

Montemanni, R., L. Gambardella, A. Rizzoli, and A. Donati (2005). Ant Colony System for a Dynamic Vehicle Routing Problem. *Journal of Combinatorial Optimization 10*(4), 327–343.

Müller, H. J. (1997). Towards Agent Systems Engineering. *Data & Knowledge Engineering 23*(3), 217–245.

Munkres, J. (1957). Algorithms for the assignment and transportation problems. *Journal of the Society for Industrial & Applied Mathematics 5*(1), 32–38.

Nalepa, J., M. Blocho, and Z. Czech (2014). Co-operation schemes for the parallel memetic algorithm. In R. Wyrzykowski, J. Dongarra, K. Karczewski, and J. Wasniewski (Eds.), *Parallel Processing and Applied Mathematics*, Volume 8384 of *Lecture Notes in Computer Science*, pp. 191–201. Springer Berlin Heidelberg.

Netzer, T., J. Krause, L. Hausmann, and N.-A. Hermann (2014). Travel, Transport, and Logistics Same day delivery: The next evolutionary step in parcel logistics. Technical report, McKinsey and Company.

Neumann, J. V. and O. Morgenstern (1944). *Theory of Games and Economic Behavior*. Princeton, NY, USA: Princeton University Press.

Padberg, M. and G. Rinaldi (1991). A Branch-and-Cut Algorithm for the Resolution of Large-Scale Symmetric Traveling Salesman Problems. *SIAM Review 33*(1), 60–100.

Pankratz, G. (2005). A Grouping Genetic Algorithm for the Pickup and Delivery Problem with Time Windows. *OR Spectrum 27*(1), 21–41.

Parragh, S., K. Doerner, and R. Hartl (2008a). A survey on pickup and delivery problems. *Journal für Betriebswirtschaft 58*(1), 21–51.

Parragh, S., K. Doerner, and R. Hartl (2008b). A survey on pickup and delivery problems. *Journal für Betriebswirtschaft 58*(2), 81–117.

Pawlaszczyk, D. and I. J. Timm (2007). A Hybrid Time Management Approach to Agent-Based Simulation. In C. Freksa, M. Kohlhase, and K. Schill (Eds.), *KI 2006: Advances in Artificial Intelligence*, Volume 4314 of *Lecture Notes in Computer Science*, pp. 374–388. Springer Berlin Heidelberg.

Perugini, D., D. Lambert, L. Sterling, and A. Pearce (2003). A Distributed Agent Approach to Global Transportation Scheduling. In *Proceedings of the IEEE/WIC International Conference on Intelligent Agent Technology (IAT)*, pp. 18–24.

Pillac, V., M. Gendreau, C. Guret, and A. L. Medaglia (2013). A review of dynamic vehicle routing problems. *European Journal of Operational Research 225*(1), 1–11.

Pokahr, A. and L. Braubach (2009). From a Research to an Industrial-Strength Agent Platform: Jadex V2. In H. R. Hansen, D. Karagiannis, and H.-G. Fill (Eds.), *Tagungsband der neunten Internationalen*

Tagung Wirtschaftsinformatik – Business Services: Konzepte, Technologien, Anwendungen (WI), pp. 769–778. Österreichische Computer Gesellschaft Wien.

Pokahr, A., L. Braubach, J. Sudeikat, W. Renz, and W. Lamersdorf (2008). Simulation and Implementation of Logistics Systems based on Agent Technology. In T. Blecker, W. Kersten, and C. Gertz (Eds.), *Proceedings of the Hamburg International Conference on Logistics 2008: Logistics Networks and Nodes*, pp. 291–308. Erich Schmidt Verlag.

Pokahr, A., L. Braubach, A. Walczak, and W. Lamersdorf (2007). Jadex - Engineering Goal-Oriented Agents. In F. Bellifemine, G. Caire, and D. Greenwood (Eds.), *Developing Multi-Agent Systems with JADE*, pp. 254–258. Chichester, UK: John Wiley & Sons.

Potvin, J. and J. Rousseau (1995). An Exchange Heuristic for Routeing Problems with Time Windows. *Journal of the Operational Research Society 46*(12), 1433–1446.

Potvin, J.-Y. (1993). State-of-the-art survey – the traveling salesman problem: A neural network perspective. *ORSA Journal on Computing 5*(4), 328–348.

Potvin, J.-Y. (1996). Genetic algorithms for the traveling salesman problem. *Annals of Operations Research 63*(3), 337–370.

Potvin, J.-Y. and S. Bengio (1996). The Vehicle Routing Problem with Time Windows Part II: Genetic Search. *INFORMS Journal on Computing 8*(2), 165–172.

Potvin, J.-Y. and J.-M. Rousseau (1993). A parallel route building algorithm for the vehicle routing and scheduling problem with time windows. *European Journal of Operational Research 66*(3), 331–340.

Psaraftis, H. N. (1980). A Dynamic Programming Solution to the Single Vehicle Many-to-Many Immediate Request Dial-a-Ride Problem. *Transportation Science 14*(2), 130–154.

Pureza, V. and G. Laporte (2008). Waiting and buffering strategies for the dynamic pickup and delivery problem with time windows. *INFOR: Information Systems and Operational Research 46*(3), 165–176.

Rao, A. S. and M. P. Georgeff (1991). Modeling Rational Agents within a BDI-Architecture. In R. Fikes and E. Sandewall (Eds.), *Proceedings of the Second International Conference On Principles of Knowledge Representation and Reasoning*, pp. 473–484. Morgan Kaufmann.

Rao, A. S. and M. P. Georgeff (1995). BDI Agents: From Theory to Practice. In *Proceedings of the First International Conference on Multi-Agent Systems (ICMAS)*, pp. 312–319. AAAI.

Renaud, J., G. Laporte, and F. F. Boctor (1996). A tabu search heuristic for the multi-depot vehicle routing problem. *Computers & Operations Research 23*(3), 229–235.

Ribeiro, G. M. and G. Laporte (2012). An adaptive large neighborhood search heuristic for the cumulative capacitated vehicle routing problem. *Computers & Operations Research 39*(3), 728 – 735.

Rochat, Y. and E. Taillard (1995). Probabilistic diversification and intensification in local search for vehicle routing. *Journal of Heuristics 1*(1), 147–167.

Ropke, S., J.-F. Cordeau, and G. Laporte (2007). Models and branch-and-cut algorithms for pickup and delivery problems with time windows. *Networks 49*(4), 258–272.

Rosenschein, J. S. and G. Zlotkin (1994). *Rules of Encounter: Designing Conventions for Automated Negotiation Among Computers*. Cambridge, MA, USA: The MIT Press.

Rosin, C. D. (2011). Nested Rollout Policy Adaptation for Monte Carlo Tree Search. In *Proceedings of the Twenty-Second International*

Joint Conference on Artificial Intelligence (IJCAI), Volume 1, pp. 649–654. AAAI Press.

Russell, S. and P. Norvig (2010). *Artificial Intelligence. A Modern Approach* (3rd ed.). Prentice Hall.

Sandholm, T. W. (1999). Distributed Rational Decision Making. In G. Weiss (Ed.), *Multiagent Systems: A Modern Approach to Distributed Artificial Intelligence* (1st ed.). Cambridge, MA, USA: The MIT Press.

Sano, Y., Y. Kadono, and N. Fukuta (2014). A Performance Optimization Support Framework for GPU-based Traffic Simulations with Negotiating Agents. In *Proceedings of Seventh International Workshop on Agent-based Complex Automated Negotiations (ACAN)*.

Savelsbergh, M. and M. Sol (1998). Drive: Dynamic routing of independent vehicles. *Operations Research 46*(4), 474–490.

Savelsbergh, M. W. P. (1992). The Vehicle Routing Problem with Time Windows: Minimizing Route Duration. *ORSA Journal on Computing 4*(2), 146–154.

Scheer, A. W. (1999). *ARIS. Vom Geschäftsprozess zum Anwendungssystem*. Springer Berlin Heidelberg.

Schillo, M., C. Kray, and K. Fischer (2002). The Eager Bidder Problem: A Fundamental Problem of DAI and Selected Solutions. In *Proceedings of the First International Joint Conference on Autonomous Agents and Multiagent Systems: Part 2 (AAMAS)*, New York, NY, USA, pp. 599–606. ACM.

Scholz-Reiter, B., K. Windt, J. Kolditz, F. Böse, T. Hildebrandt, T. Philipp, and H. Höhns (2004). New Concepts of Modelling and Evaluating Autonomous Logistic Processes. In *Proceedings of the Manufacturing, Modelling, Management and Control*, IFAC.

Schuldt, A. (2011). *Multiagent Coordination Enabling Autonomous Logistics*. Springer Berlin Heidelberg.

Schuldt, A., K. A. Hribernik, J. D. Gehrke, K.-D. Thoben, and O. Herzog (2010). Cloud Computing for Autonomous Control in Logistics. In K.-P. Fähnrich and B. Franczyk (Eds.), *Proceedings of the Fortieth Annual Conference of the German Society for Computer Science*, Volume 1 of *Lecture Notes in Informatics*, Leipzig, Germany, pp. 305–310. Gesellschaft für Informatik.

Searle, J. R. (1969). *Speech Act Theory*. Cambridge, MA, USA: Cambridge University Press.

Shaw, P. (1998). Using Constraint Programming and Local Search Methods to Solve Vehicle Routing Problems. In M. Maher and J.-F. Puget (Eds.), *Principles and Practice of Constraint Programming*, Volume 1520 of *Lecture Notes in Computer Science*, pp. 417–431. Springer Berlin Heidelberg.

Shi, X., Y. Liang, H. Lee, C. Lu, and Q. Wang (2007). Particle swarm optimization-based algorithms for TSP and generalized TSP . *Information Processing Letters 103*(5), 169–176.

Shoham, Y. and K. Leyton-Brown (2009). *Multiagent Systems: Algorithmic, Game-Theoretic, and Logical Foundations*. Cambridge, MA, USA: Cambridge University Press.

Sinnott, R. W. (1984). Virtues of the Haversine. *Sky and Telescope 68*(2).

Skobelev, P. (2011). Multi-Agent Systems for Real Time Resource Allocation, Scheduling, Optimization and Controlling: Industrial Applications. In V. Maík, P. Vrba, and P. Leitão (Eds.), *Holonic and Multi-Agent Systems for Manufacturing*, Volume 6867 of *Lecture Notes in Computer Science*, pp. 1–14. Springer Berlin Heidelberg.

Smith, R. G. (1977). The Contract Net: A Formalism for the Control of Distributed Problem Solving. In R. Reddy (Ed.), *Proceedings of the Fifth International Joint Conference on Artificial Intelligence (IJCAI)*, Cambridge, MA, USA, pp. 472. Morgan Kaufmann Publishers.

Smith, R. G. (1980). Communication and Control in a Distributed Problem Solver. *IEEE Transactions On Computers C29*(12), 1104–1113.

Snyder, L. V. and M. S. Daskin (2006). A random-key genetic algorithm for the generalized traveling salesman problem. *European Journal of Operational Research 174*(1), 38–53.

Solomon, M. (1987). Algorithms for the Vehicle Routing and Scheduling Problems with Time Window Constraints. *Operations Research 35*, 254–265.

Solomon, M. M. and J. Desrosiers (1988). Survey Paper – Time Window Constrained Routing and Scheduling Problems. *Transportation Science 22*(1), 1–13.

Srour, F. J. and S. van de Velde (2013). Are Stacker Crane Problems easy? A statistical study . *Computers & Operations Research 40*(3), 674–690. Transport Scheduling.

System Alliance GmbH (2014). System Alliance: Die Stückgutprofis. http://www.systemalliance.de/pdf/systemalliance_ib_3-2014.pdf(cited: 1.9.15).

Taillard, . E., P. Badeau, M. Gendreau, F. Guertin, and J.-Y. Potvin (1997). A Tabu Search Heuristic for the Vehicle Routing Problem with Soft Time Windows. *Transportation Science 31*(2), 170–186.

Tan, K., T. Lee, K. Ou, and L. Lee (2001). A messy genetic algorithm for the vehicle routing problem with time window constraints. In *Proceedings of the Congress on Evolutionary Computation*, Volume 1, pp. 679–686.

Ten Hompel, M. (2006). Zellulare Fördertechnik. *eLogistics Journal*.

Ten Hompel, M. (2010). Individuell bewegen – Das Internet der Dinge und Dienste. In W. Delfmann and T. Wimmer (Eds.), *Strukturwandel in der Logistik*, pp. 286–295. DVV Media Group.

Ten Hompel, M. (Ed.) (2012). *IT in der Logistik: Trends des Logistik-IT-Marktes auf einen Blick – vom Supply Chain Management bis zum Warehouse Management*. DVV Media Group.

Ten Hompel, M. and M. Henke (2014). Logistik 4.0. In T. Bauernhansl, M. Ten Hompel, and B. Vogel-Heuser (Eds.), *Industrie 4.0 in Produktion, Automatisierung und Logistik*, pp. 615–624. Springer Fachmedien Wiesbaden.

Thangiah, S. R., O. Shmygelska, and W. Mennell (2001). An Agent Architecture for Vehicle Routing Problems. In *Proceedings of the ACM Symposium on Applied Computing (SAC)*, New York, NY, USA, pp. 517–521. ACM.

Thomas, B. W. (2007). Waiting Strategies for Anticipating Service Requests from Known Customer Locations. *Transportation Science 41*(3), 319–331.

Toth, P. and D. Vigo (1997). Heuristic Algorithms for the Handicapped Persons Transportation Problem. *Transportation Science 31*(1), 60–71.

Vahrenkamp, R. (2007). *Logistik: Management und Strategien* (6th ed.). München, Germany: Oldenbourg.

van Lon, R. and T. Holvoet (2013, April). Evolved multi-agent systems and thorough evaluation are necessary for scalable logistics. In *IEEE Workshop on Computational Intelligence in Production And Logistics Systems*, CLIPS, pp. 48–53.

van Lon, R. R., T. Holvoet, G. Vanden Berghe, T. Wenseleers, and J. Branke (2012). Evolutionary Synthesis of Multi-agent Systems for Dynamic Dial-a-ride Problems. In *Proceedings of the Fourteenth Annual Conference Companion on Genetic and Evolutionary Computation (GECCO)*, New York, NY, USA, pp. 331–336. ACM.

Vickrey, W. (1961). Counterspeculation, Auctions, and Competitive Sealed Tenders. *Journal of Finance 16*(1), 8–37.

Vokřínek, J., A. Komenda, and M. Pěchouček (2010). Agents Towards Vehicle Routing Problems. In *Proceedings of the Ninth International Conference on Autonomous Agents and Multiagent Systems (AAMAS)*, Volume 1, Richland, SC, pp. 773–780. International Foundation for Autonomous Agents and Multiagent Systems.

Warden, T., R. Porzel, J. D. Gehrke, O. Herzog, H. Langer, and R. Malaka (2010). Towards Ontology-Based Multiagent Simulations: The Plasma Approach. In A. Bargiela, S. Azam Ali, D. Crowley, and E. J. Kerckhoffs (Eds.), *Proceedings of the European Conference on Modelling and Simulation (ECMS)*, pp. 50–56.

Weddewer, M. (2007). *Verrechnungspreissysteme für horizontale Speditionsnetzwerke*. Deutscher Universitätsverlag.

Weiss, G. (Ed.) (2013). *Multiagent Systems* (2ed ed.). Cambridge, MA, USA: The MIT Press.

Windt, K. (2008). Ermittlung des angemessenen Selbststeuerungsgrades in der Logistik - Grenzen der Selbststeuerung. In P. Nyhuis (Ed.), *Beiträge zu einer Theorie der Logistik*, pp. 349–372. Springer Berlin Heidelberg.

Wooldridge, M. (2009). *An Introduction to Multiagent Systems* (3rd ed.). Glasgow, UK: John Wiley & Son.

Wooldridge, M. (2013). Intelligent Agents. In G. Weiss (Ed.), *Multiagent Systems* (2ed ed.)., pp. 3–50. Cambridge, MA, USA: The MIT Press.

Yan, X., P. Diaconis, P. Rusmevichientong, and B. V. Roy (2004). Solitaire: Man Versus Machine. In L. K. Saul, Y. Weiss, and L. Bottou (Eds.), *Advances in Neural Information Processing Systems 17*, pp. 1553–1560. Cambridge, MA, USA: The MIT Press.

Yang, J., P. Jaillet, and H. Mahmassani (2004). Real-Time Multivehicle Truckload Pickup and Delivery Problems. *Transportation Science 38*(2), 135–148.

Zhang, W. and R. E. Korf (1995). Performance of linear-space search algorithms. *Artificial Intelligence 79*(2), 241–292.

Zhenggang, D., C. Linning, and Z. Li (2009). Improved Multi-Agent System for the Vehicle Routing Problem with Time Windows. *Tsinghua Science and Technology 14*(3), 407–412.

Printed in the United States
by Bookmasters

Printed in the United States
By Bookmasters